# Chemicals for the Automotive Industry

Edited by

**J. A. G. Drake**

THE ROYAL SOCIETY OF CHEMISTRY
1841-1991

Proceedings of a symposium organized by the North East Region of the Industrial Division of the Royal Society of Chemistry, University of York, 11th–12th September 1990.

Special Publication No. 93

ISBN 0-85186-427-9

A catalogue record of this book is available from the British Library



Published by The Royal Society of Chemistry,
Thomas Graham House, Science Park, Cambridge CB4 4WF

Printed and bound in Great Britain by J. W. Arrowsmith Ltd, Bristol, England

# PREFACE

When the committee met eighteen months ago, several topics were considered for a symposium,but it was the theme "Chemicals and Materials for the Automotive Industry" which it was felt would be the most appropriate. From the attendance and list of speakers, a good choice was made.

The factors that influenced the committee were the importance of the motor car/truck business. It consumes about 30% of the world's crude oil (almost all used as transport fuel; about 40% if the manufacture of the automobile is also included) and up to 80% of the world's rubber (natural and synthetic) for tyres (over one billion tyres are made every year). Since about 50 million cars/trucks are made every year, a 1 kg material use per car yields 50,000 tonnes. The current figure is now about 60 kg plastic use per car, not counting tyres, giving an annual use of about 3 million tonnes. After-care of automobiles is another large consumer of chemicals.

The car industry is very conservative, probably because of safety and the threat of recalls. It takes several years for a manufacturer to live down the stigma of a bad reputation so they are cautious towards innovation. Although illustrations of the "total concept car" and "car of the future" have been seen for the past 25 years they still have not appeared. Despite this, progress has been and still is being made on all fronts, many of which are a combination of chemicals/materials and engineering, for example tyres, fuel efficiency, corrosion prevention, use of plastics for bumpers, petrol tanks etc. An example of just how much progress has been made can be illustrated by considering tyres. In 1910 (22 years after Dunlop's invention of the rubber pneumatic tyre), the cost of tyres was the equivalent of three weeks wages, and an inner tube one weeks wage. Every visit to a garage for petrol also included inflation of tyres (butyl rubber was not discovered until 1937). The average consumption was 37.5 tyres per year; the average journey, approximately 100 to 200 miles, required five spare tyres. Even if left in the garage and not used, the tyres degraded in six months (amine anti-oxidants were not introduced until the 1920s). It must be admitted, progress has been made when one considers that 30 to 50 000 miles per tyre is common and the average motorist travels 100 000 miles without a puncture.

Corrosion is another area of advancement, with extended warranties approaching ten years. Gone are the days when tell-tale signs of rust appeared months after the twelve month warranty had expired.

Some of the reasons how and why progress has been achieved include, making the car more fuel efficient, that is a lighter weight; generally replacing metal by plastics. Cars are made cheaper by more efficient manufacture, for example car seats are now made in two operations compared with armies of upholsterers a few years ago. Legislation has enforced changes for example fuel consumption, in the USA must not drop too low for the size of the vehicle. Fuel composition (non-leaded), and exhaust emissions (the use of catalytic converters) have strict limits. Safety legislation is, for example, changing the design of fuel tanks, which are now made from impact resistant high density polyethylene and look like a three dimensional model of a human heart. Many materials developed for aerospace applications are finding use in under-the-bonnet parts in automobiles. Some aeroplanes made out of titanium/aluminium are glued together - will adhesives play a more important role in the future motor car? Finally, grand prix racing is introducing many new technologies, especially on safety.

Finally, the programme. The committee grouped the topics under several themes. Many of the speakers talked about future materials and we hope that these are successfully introduced into the automotive industry. The majority of the lectures are included in this volume.

In addition to advice, help and encouragement from many members of the Royal Society of Chemistry, the symposium was held under the auspicies of the Society. Sponsorship was also received from BASF plc, Seal Sands and ICI Chemicals and Polymers, Wilton. However, success was only possible through the goodwill and hard work of a small organising committee *viz* Eric Lee (BASF), Dr Jack Candlin (ICI), Graham Henney, Keith Hardy and Dr John Drake.

Dr J.P. Candlin,
Dr J.A.G. Drake, April 1991

## CONTENTS

# TRENDS IN THE AUTOMOTIVE INDUSTRY AND THEIR EFFECT ON MATERIALS REQUIREMENTS

Dr M.G. Lofthouse

Du Pont (UK) Ltd
Hemel Hempstead
Hertfordshire, HP2 7DP
United Kingdom

## 1 THE CAR OF THE FUTURE

Some years ago, Du Pont ran an advertisement aiming to show what the car of the future would look like. However, in the way that one does with advertising, we hid the vehicle with a dust cover. I can now provide a glimpse of what was underneath.

Since car design is dictated by peoples' preferences, the car must have style: Rolls Royce or Bentley would do nicely. In true macho image, it should be powerful, like a dragster. Inside, it should have all the comforts of the home. Running costs should be as low as possible: put all the mechanical parts on the outside, and it becomes easy to maintain! The manufacturers would like to have a say, so it should be built by robots. Some of these needs appear to be mutually exclusive. But with new technologies and new materials, progress is being achieved on many fronts. I will show what I mean with selected examples in four areas:

### 1.1 Product engineering

The first stages in car production involve design using computers, and selecting the correct material; one which will meet the needs of the part and be capable of being fabricated economically.

Everyone's idea of a nice-looking car is the Porsche Carerra! A look at the front and rear spoiler confirms how design has come a long way. Today, finite element models allow a closer match of material properties which the end use forces on the part. In the case of the Carerra, a flexible polyester was selected. Using material property data, the flow of the plastic through the sprues and runners into the cavity can be modelled, and design changes made to remove weld lines from areas of high stress.

Only after these models are fully analysed will the car companies be willing to spend the £200 000 needed to make the mould. Note also that the part comes out of the mould using pick and place robots, both to help control costs and to avoid distortion of the parts. Exterior parts are often close to painted metal and need to be painted to give a good match, hence the need for good paint adhesion. A great deal of work has been done to ensure that painted plastic parts have excellent surface finish to comply with the exacting standards in the car industry.

## 1.2 Quality and cost

Quality and reliability are obviously increasingly important, and quality systems are widely used to ensure reproducible results; the car manufacturers' standards are demanding. In plastics, elastomers, fibres and coatings, BS 5750 or ISO 9000 have become the accepted standards, and closely parallel the development of the car manufacturers' own systems such as FORD's Q1 and Q101.

No discussion in the car industry would be complete without some mention of costs! Plastics provide many benefits to the car companies as well as motorists. These include: corrosion resistance, high strength/weight ratio, colourability, transparency, good insulators and design flexibility. However, unless they can provide cost effectiveness, they would not be used in the increasing quantities we see today. Even so there are many potentially exciting applications that will not be realised unless engineers and designers can find more reliable ways to check the match of a part's performance to the correct material choice. Prototyping is an established method of doing this and for some time the search has been on to find speedier ways of doing it.

In our laboratories in Geneva, we have a SOMOS (solid modelling system) machine, sometimes called stereolithography. A computer-modelled design is scanned and the output controls a mirror which directs lights from a laser onto a bath containing a photopolymer. As the layers polymerise, the support platform moves downwards bringing fresh monomer to the surface. When the model is fully scanned, the platform is raised and the part can be further treated. This direct link between the computer model and a solid object allows a significant cost saving by reducing the time for product development. Examples of components we have modelled with SOMOS are a constant velocity joint using a flexible photomonomer, and a simple gear train in several parts. Currently, we are working to extend the useful properties of the photopolymers to move this technique into the direct fabrication of parts.

## 1.3 Major trends

Three trends in particular have an effect on the choice and use of materials:

Safety. One of the first automotive safety features in which Du Pont was involved was laminated glass windscreens. The material used was polyvinylbutyraldehyde (PVB, trade name BUTACITE™) and sheets of this plastic are laminated between two pieces of glass to form the typical laminated structure seen widely today. An unrestrained body in a crashing car is an awesome sight, and without laminated glass the body could fly through the windscreen, suffering fatal injuries. Even with laminated glass, the face can suffer because of splinters from the internal layer of glass. While the construction of windscreens has hardly changed in 25 years, recent advances in the use of polyester films on glass offer a way round this problem of abrasion trauma.

Such trauma injuries are less likely if the body can be restrained. Hence the development of so-called passive restraints such as seat belts which wrap around you on entering the car. They incorporate oriented polyester film in the drive mechanisms.

Air bags are a recent innovation. Introduced in high-cost cars initially, they are now being offered on mid-range models. The bags, made of nylon fibre sometimes with a rubber coating, are explosively inflated in the first microseconds of a crash. The bag has to emerge from behind the steering wheel or glove box through a specially designed air bag door, made from a plastic material which tears without leaving sharp edges to rupture the bag. Material development here is somewhat sensitive owing to the critical safety application. Air bags are expected to find their way into all parts of the passenger compartments, and are seen as one way to reduce injuries from side impacts.

Comfort. Many comfort features in today's cars, and increasingly, those in the future, will be electronically driven. The value of electronics in cars will double in the next few years, to an estimated £500 per vehicle. Some of these applications, and this is not an exhaustive list, include engine control, active suspension, RFI filtration, distance control and navigation aids using compact discs and satellite communication!

Higher temperatures. The area of the car where the biggest increase in temperature has occurred is the engine compartment. Factors leading to this include the changing shape of the front end of cars to the V-shape to improve aerodynamics, the tighter packing of parts under the bonnet, air dams and fascias reducing air flow through the engine compartment, and catalytic converters and turbo-chargers. The result has been an increase in temperature from 90 C to 130 C over the past decade, with 150 C expected before too long!

Despite such temperatures, polymer applications flourish. One example is the oil pick-up in ZYTEL™ (Du Pont's glass reinforced nylon) on the Fiat FIRE (Fully Integrated Robotised Engine) engine. This design consolidated eight parts down to just two, achieving weight reduction as well as assembly simplification. Another example is a rocker cover in MINLON™, Du Pont's mineral reinforced nylon, used by Nissan and in Rover's new K series engine. The benefits, in addition to part consolidation, have been weight reduction and a decrease in noise of 1 to 3 decibels. Similarly, high-temperature performance rubbers such as ethylene acrylics (VAMAC™) are increasingly used to overcome problems with traditional cork gaskets.

Warranty savings. If the engine has to be stripped down because of an oil leak, the warranty costs could be huge. VITON™ Du Pont's high performance fluoro-elastomer is much more expensive than traditional nitriles, but the cost saving from lower warranty costs from failure of crankshaft seals far outweighs this. The increase in temperatures also means that more aggressive antioxidants and detergents are needed, additional reasons for the use of the fluoro-elastomers to ensure rotary seal performance.

The cooling system also has to cope with the increased heat level. Reinforcing fibres in radiator hoses such as rayon, nylon and polyester are giving way to KEVLAR™ aramid fibre. While it is perhaps better known in bullet-proof vests, some 6 to 10 million hoses are now made with KEVLAR™ reinforcement.

Reducing noise. Noise and vibrations (knocks, squeaks and rattles) are the things all of us as motorists complain about most. Not surprisingly, all car makers have NVH (Noise, Vibration and Harshness) groups to deal with this problem. Over the years, many materials have been developed to control noise. For example, bushings made of Du Pont's HYTREL™ polyester elastomer can help absorb noise from anti-roll bar systems. Plastic (mineral reinforced nylon) window guides reduce glass rattle in doors. Metal parts overmoulded with HYTREL™ polyester elastomer eliminate rattles from locks.

## 1.4 Environmental issues

This is perhaps the single biggest area of concern today. Leaving aside for the moment pollution from noise, cars pollute during their production stage, in everyday use, and at the end of their useful life.

Emissions. Carbon dioxide, nitrogen oxides, carbon monoxide and hydrocarbons are emitted in large volumes and contribute to today's health problems and tomorrow's global warming or greenhouse effect. Catalytically converting nitrogen oxides and carbon monoxide to carbon dioxide and nitrogen helps to reduce acid emissions, but increases warming gases. The only way to reduce these effects is to build smaller, lighter cars with more efficient, lean burn engines. Alternative fuels such as hydrogen may be used in the future.

Emissions from processing and waste can also be harmful. Examples include chlorofluorocarbon blowing agents for foamed plastics, and chlorofluorocarbons coming from scrapped air conditioning units. Damage to the ozone level has at least been recognised and alternatives such as 1,3,4-hydrofluorocarbons are being scaled up to replace the chlorofluorocarbons.

Waste. Plastic waste from scrapped cars will increase from one million tonnes today to at least two million tonnes in the next ten years. The problems of emission and waste are closely linked. One way to reduce emissions is to lower the weight of the car. This can be achieved most effectively by changing metal parts to plastic, which can result in a minimum 20% weight saving.

This has certainly been recognised in the USA, where car design is strongly influenced by CAFE (Corporate Automotive Fuel Efficiency). This is basically a miles per gallon target that *all* new cars must meet for a given engine size. It is affected most by weight, hence the driving force to convert large parts such as the air intake manifold into plastic. This will be made of a 35% glass reinforced ZYTEL™ resin (from Du Pont) using a lost core technique whereby cores made from a eutectic metal are first moulded into the inner shape of the manifold then overmoulded with plastic and finally the core melted to recover the metal.

Along with other suppliers, Du Pont recognises that plastics waste *is* an important issue for the automotive industry today. But it must be put into perspective versus other sources of plastics waste, for example packaging.

Today, transport products account for 7% of the solid waste stream, compared to 35% by packaging products. It should also be noted that so far as *automotive* waste is concerned, plastics amount to 10% versus 68% for steel.

The plastics industry, car makers and various Institutes and Federations have agreed on a four-point strategy for dealing with waste:

a) minimisation at source by efficient design and controlled manufacturing technologies,
b) recycling, where the value of the material can be utilised in lower performance parts,
c) energy recovery (plastics are an excellent source of energy), and
d) landfill of irreducible residues.

This provides a coherent and comprehensive approach for meeting the public's expectations and needs in this area.

## 2 CONCLUSION

Du Pont and its associated companies provide materials for all parts of the car: from paint to plastic, rubber to fibres and electronics, and through Conoco, to the oils and fuels used by motorists. Thus, we are closely involved at every stage in a car's life cycle: from design to production engineering, through the vehicle's daily usage to after it has been scrapped. That involvement makes us acutely aware of the contribution, and the opportunities which new materials offer the automotive industry in responding to today's and tomorrow's issues.

# TECHNICAL INFORMATION ON TYRE SIDEWALLS

D. Trigg

Avon Tyres Ltd
Melksham
Wiltshire, SN12 8AA
United Kingdom

## 1 INTRODUCTION

During the 1960's in the USA Ralph Nader and his team investigated on behalf of the Federal Department of Transportation the production of cars which were "unsafe at any speed". This was in accordance with Federal Motor Vehicle Safety Standard FMVSS 109. From the investigation car tyres had to be identifiable by country of origin and marked with the following information to ensure tyre suitability and safety.

a) date of manufacture with an identity code AT.....
b) tread wear grading
c) temperature grading
d) traction
e) tyre approval number

This requirement is known as Uniform Tyre Quality Grading (U.T.Q.G.).

With the Common Market formation these ideas were "imported" into Europe and the United Kingdom and similar requirements were considered. The first implementation directive from Brussels was ECE 30 regarding car tyres and from the middle of 1985 all new car tyres had to be marked. Numerical codes were given for each country of origin; the United Kingdom was allocated E11. This applied across the range of car tyres except for "C" type and V/Z speed rated car tyres. As from that date it became illegal to sell new tyres which were not "E" marked and tested to ECE 30 standards.

Car Tyre Sidewall Markings - Example

155R12 76S. The 155 figure indicates the nominal section width of tyre when fitted as 155 mm. "R" in the size marking denotes a Radial construction casing with textile or steel tread belts. "12" is the nominal wheel diameter to which the tyre should be fitted, i.e. 12 inch diameter. "76" is the Load Index figure and from the Load Tables gives a maximum load for this tyre of 400 kg at the specified pressure. "S" is the Speed Symbol showing the maximum speed for tyre operation in this case 113 m.p.h. (180 km $h^{-1}$).

Radial. Denotes the type of tyre construction. There are other constructions such as cross (diagonal ply) and bias belted tyres. The latter type was used in the USA but some high speed motorcycle tyres are produced in this country at the present time.

Tubeless. The tyre has to be identified as Tubeless or Tube Type construction so that correct fitment can be made for the wheel type on a vehicle.

Treadwear 160, Traction A and Temperature B. These markings are to conform with USA FMVSS 109 requirements. Test tyres are produced in the USA and checked for treadwear, traction and running temperature. Tyres imported into the USA must conform with these standards and be graded accordingly. Treadwear 160 confirms a mileage 60% higher than the basic FMVSS 109 requirements. Traction capability is graded A, B or C, therefore "A" is confirmation of maximum grip potential against the test standard. Tyre running temperature is also graded as A, B or C, temperature "B" is within the permitted test variance .

Made in Great Britain. The country of origin has to be stated so that if a recall of defective tyres has to be made the tyres can be clearly identified.

North American Department of Transportation (DOT) Compliance Symbol e.g. DOT ATC8 AUA108. The letters AT confirm the registration of Avon Tyres and the following letters and digits identify the tyre as an Avon 155R12 76S Turbosteel, Tubeless, Radial tyre. To meet Federal requirements the date of manufacture has to be shown for possible recall purposes. The code is based on a ten year cycle, the first two digits indicate the calendar week "10" and the figure "8" indicates 1988. The system was adopted, generaly, late in 1970 but, unfortunately, the committee which agreed the system did not appreciate that tyres could overlap the ten year cycle. This meant that if a particular size and type of tyre was in production for a long period of time it was not easy to confirm when the tyre was produced, e.g. 1970, 1980 or 1990. Two methods of overcoming this problem have been adopted:-

a) The age of tyres prior to 1979 could easily be established as all Tyre Companies had their own date code marking. 1980 was the nominal starting point of the system and so this raised no problems.

b) Since, from 1990 onwards the dates would repeat the sequence, it would be impossible to tell whether a tyre was ten years or only weeks old. The answer for the 1990 cycle was to use a triangular mark after the third figure of the date code.

155SR12 RADIAL. An interim tyre identification showing speed rating "S" in the early styling, plus construction.

M + S. A marking to indicate that the tyre has a winter type tread pattern, M + S means mud and snow use.

E11 02409. ECE 30 type approval mark and certificate number. The tyre has to pass strength, load and endurance testing before the approval is given.

Re-inforced. Some car tyres are re-inforced for extra load capacity for use on estate cars or light van derivatives and this must be clearly identifiable on the tyres.

Tyre construction details. For the USA market the tyre construction must be confirmed. Thus, in this instance, there are two casing plies of rayon material. Under the tread area there are two rayon plies (casing) plus two steel plies (tread belts).

Avon Turbosteel. The manufacturer's name, or brand name, must be clearly identified.

Load and pressure marking. For FMVSS test purposes a maximum load and pressure must be shown. This often causes confusion in the home market as customers misunderstand the purpose of the figures which do not always follow and conform to United Kingdom practice.

T.W.I. This mark does not always appear on tyres but it means "Tread Wear Indicator". There are six equally spaced mouldings at the base of the tread pattern at 1.6 mm remaining pattern depth. Thus, when these wear "bars" are level with the tread it was intended that the tyre should be removed and replaced. Such a tyre would be the subject of prosecution if used beyond that degree of tread wear. However, under the United Kingdom tyre regulations (regulation 27), a 1 mm minimum pattern depth is permitted, providing it is for the full circumference of the tread, measuring at least three-quarters of the tread width for the full circumference and with "clearly visible" pattern in the remaining quarter of the tread width. The difficulties in applying such requirements are all too easy to appreciate. It should not be too long before the minimum 1.6 mm depth for the full circumference and width of the tread pattern becomes legally applicable.

It was later recommended that similar checks should be made on truck tyres and , at that time, the majority of tyres had a maximum speed capability of 100 km $h^{-1}$ (62 m.p.h.). As trucks and coaches became faster, with higher sustained speed capability, it was felt that the margin of safety was too low and ECE 54 type approval standards were laid down. From the first of April 1987, tyres have been progressively marked and from the first of October 1990 truck tyres have also to be "type approved" with load/speed indices.

As far as DOT Test Centres are concerned New Plated Tables (permitted maximum axle loadings) are being ratified for several reasons.

a) Historically, Plated Axle Loads were higher than those shown in tyre manufacturer's data, for example 11R22.5 was 6 100 kg front axle load in Plated Tables. Manufacturer's tables gave 6 000 kg and on that basis the tyres could be regarded as unsuitable for the operation.

b) Tyres have progressively been given a load/speed index marking e.g. 146/143 that is front axle 6 000 kg, rear axle 10 900 kg.

c) Because of the discrepancy between Load Index figures and Plated Loads, the Load Index figures have been raised to conform. Therefore, if the original 11R22.5 index was 146/143, to conform to plating requirements this would now be 148/144 giving a front axle 3 150 x 2 = 6 300 kg and rear axle 2 800 x 4 = 11 200 kg load. Avon have progressively raised the tyre loading capability to meet the new standards prior to the implementation of the new regulation so that the majority of operators will have no problems unless new tyres are fitted which do not have "E" markings. (Note, to increase the load/speed index means a full scale retest of the tyres to obtain the type approval number as regards performance and this approval has to be substantiated by test data.)

d) Some tyres have a "unique point" marking, such as the 15R22.5/385.65R22.5 size, that is 160J/158K. Therefore at J speed 100km $h^{-1}$ (62 m.p.h.) the load per tyre is 4 500 kg (130 p.s.i. and 9 000 kg per axle), while at K speed 110 km $h^{-1}$ (68 m.p.h.) the load per tyre is reduced to 4 250 kg (125 p.s.i. and 8 500 kg per axle).

e) Speed Index. To conform with speed standards tyres are now marked with a Speed Index Letter, for example "L" - 120 km $h^{-1}$ (75 m.p.h.) or "M" - 130 km $h^{-1}$ (81 m.p.h.).

f) The original Plated Tables indicated a size and ply rating; theoretically this should now be replaced by the Load Index figures. However, some Test Centres still insist on a ply rating until the new tables are in general use. Avon have kept a ply rating figure as a "belt and braces" confirmation.

Truck Tyre Sidewall Markings - Example

Tyre Size Marking - 11.00R20. This marking indicates a nominal section width, when fitted to the correct rim width of 11.00 inches. The letter "R" denotes radial construction and "20" indicates the rim diameter of 20 inches. This size marking also denotes a tube type tyre. ER20 was an earlier alpha numeric marking. Later tubeless tyres are fitted to 22.5 inch diameter rims and the marking then would be 12R22.5 (ER22.5).

Load Speed Index - 149/145K. For front axle (steer) loading the first figure is used. Thus 149 gives a front, single, tyre loading of 3 250 kg or an axle loading of 6 500 kg. For dual (twin) rear tyre fitment the second figure "145" applies. Therefore each tyre gives a capacity of 2 900 kg or 11 600 kg with dual fitment on the axle.

The Speed Index "K". This confirms a maximum speed of 110 km $h^{-1}$ (68 m.p.h.).

DOT AT2L A8D 018. The DOT marking confirms that the tyre conforms to FMVSS standards, "AT2L A8D" is confirmation that the tyre was produced by Avon Tyres and is a 11.00R20 149/145K Radial tyre of Steermaster type. "018" is the date of production, the first week in January 1988. As for car tyres the 1990 date will have a triangular marking after the third figure to denote the decade.

Tube Type. The tyre is marked with the construction type, tube type or tubeless.

Regroovable. Certain truck tyres are designed to have sufficient undertread so that when the tread pattern depth is worn to the legal minimum depth it can be regrooved (recut) in order to obtain additional mileage. Specific instructions have to be provided for the maximum depth and width of regrooving and the tread pattern to be adopted.

Specific Fitment. The example quoted is designed primarily for front (steer) wheel use and although the casing is normal the tread pattern is of asymmetric design to resist outer shoulder wear which is common on truck operation. These tyres are fitted according to the instruction to ensure correct application.

Tyre Construction. Sidewall 1 ply steel and tread 5 ply steel have a single casing ply construction of equivalent strength to early cross ply 18 ply configuration. The tread plies take into account the single casing ply, thus there are four additional steel belts to stiffen the tread area to give the tyre the correct cornering power, stiffness and wear characteristics. In the USA ply rating (casing strength index) is shown by letter code, thus "A" represents a two ply rated tyre and for a 16 ply rated tyre the letter "H" applies.

Avon. Manufacturer's name.

Made in England. Country of origin in case of recall for defects.

Steermaster. Commercial name of identity. Some companies use numerical codes to identify tyre types such as "G291" or "AR170".

AGR 6C. Tread pattern identification number.

1174M. Individual tyre mould number.

Load and pressure marking (not applicable in the United Kingdom). For North American requirements tyre pressures and speeds have to be shown together with maximum speed.

Tyre identification. Each tyre has a serial number for individual identification, for example "19032578". In the case of Avon Tyres the first three figures "190" are the date code, calendar week 19 of 1990. The remaining six digits give the individual serial number of each specific tyre.

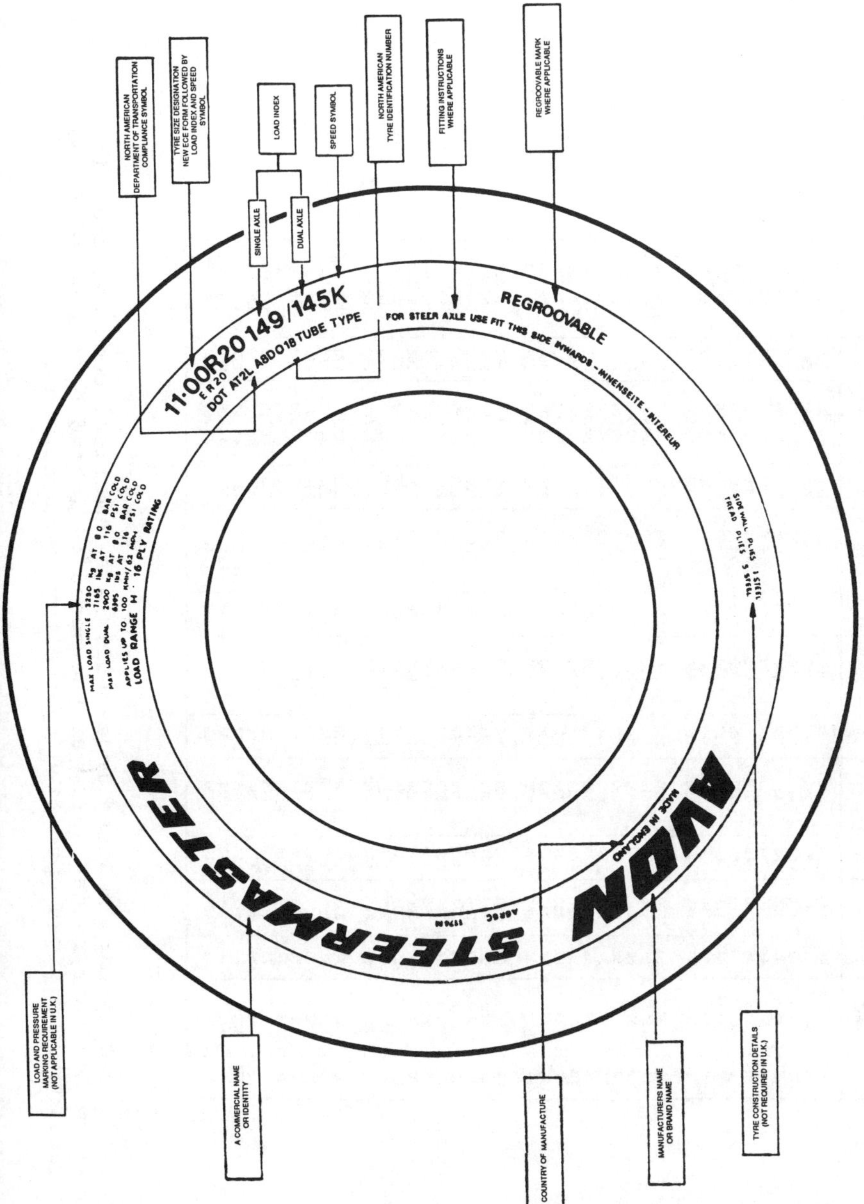


Appendix 1 Tyre Markings

| LI | kg | LI | kg | LI | kg | LI | kg | LI | kg | LI | kg | LI | kg |
|---|---|---|---|---|---|---|---|---|---|---|---|---|---|
| 0 | 45 | 40 | 140 | 80 | 450 | 120 | 1 400 | 160 | 4 500 | 200 | 14 000 | 240 | 45 000 |
| 1 | 46.2 | 41 | 145 | 81 | 462 | 121 | 1 450 | 161 | 4 625 | 201 | 14 500 | 241 | 46 250 |
| 2 | 47.5 | 42 | 150 | 82 | 475 | 122 | 1 500 | 162 | 4 750 | 202 | 15 000 | 242 | 47 500 |
| 3 | 48.7 | 43 | 155 | 83 | 487 | 123 | 1 550 | 163 | 4 875 | 203 | 15 500 | 243 | 48 750 |
| 4 | 50 | 44 | 160 | 84 | 500 | 124 | 1 600 | 164 | 5 000 | 204 | 16 000 | 244 | 50 000 |
| 5 | 51.5 | 45 | 165 | 85 | 515 | 125 | 1 650 | 165 | 5 150 | 205 | 16 500 | 245 | 51 500 |
| 6 | 53 | 46 | 170 | 86 | 530 | 126 | 1 700 | 166 | 5 300 | 206 | 17 000 | 246 | 53 000 |
| 7 | 54 5 | 47 | 175 | 87 | 545 | 127 | 1 750 | 167 | 5 450 | 207 | 17 500 | 247 | 54 500 |
| 8 | 56 | 48 | 180 | 88 | 560 | 128 | 1 800 | 168 | 5 600 | 208 | 18 000 | 248 | 56 000 |
| 9 | 58 | 49 | 185 | 89 | 580 | 129 | 1 850 | 169 | 5 800 | 209 | 18 500 | 249 | 58 000 |
| 10 | 60 | 50 | 190 | 90 | 600 | 130 | 1 900 | 170 | 6 000 | 210 | 19 000 | 250 | 60 000 |
| 11 | 61.5 | 51 | 195 | 91 | 615 | 131 | 1 950 | 171 | 6 150 | 211 | 19 500 | 251 | 61 500 |
| 12 | 63 | 52 | 200 | 92 | 630 | 132 | 2 000 | 172 | 6 300 | 212 | 20 000 | 252 | 63 000 |
| 13 | 65 | 53 | 206 | 93 | 650 | 133 | 2 060 | 173 | 6 500 | 213 | 20 600 | 253 | 65 000 |
| 14 | 67 | 54 | 212 | 94 | 670 | 134 | 2 120 | 174 | 6 700 | 214 | 21 200 | 254 | 67 000 |
| 15 | 69 | 55 | 218 | 95 | 690 | 135 | 2 180 | 175 | 6 900 | 215 | 21 800 | 255 | 69 000 |
| 16 | 71 | 56 | 224 | 96 | 710 | 136 | 2 240 | 176 | 7 100 | 216 | 22 400 | 256 | 71 000 |
| 17 | 73 | 57 | 230 | 97 | 730 | 137 | 2 300 | 177 | 7 300 | 217 | 23 000 | 257 | 73 000 |
| 18 | 75 | 58 | 236 | 98 | 750 | 138 | 2 360 | 178 | 7 500 | 218 | 23 600 | 258 | 75 000 |
| 19 | 77 5 | 59 | 243 | 99 | 775 | 139 | 2 430 | 179 | 7 750 | 219 | 24 300 | 259 | 77 500 |
| 20 | 80 | 60 | 250 | 100 | 800 | 140 | 2 500 | 180 | 8 000 | 220 | 25 000 | 260 | 80 000 |
| 21 | 82.5 | 61 | 257 | 101 | 825 | 141 | 2 575 | 181 | 8 250 | 221 | 25 750 | 261 | 82 500 |
| 22 | 85 | 62 | 265 | 102 | 850 | 142 | 2 650 | 182 | 8 500 | 222 | 26 500 | 262 | 85 000 |
| 23 | 87.5 | 63 | 272 | 103 | 875 | 143 | 2 725 | 183 | 8 750 | 223 | 27 250 | 263 | 87 500 |
| 24 | 90 | 64 | 280 | 104 | 900 | 144 | 2 800 | 184 | 9 000 | 224 | 28 000 | 264 | 90 000 |
| 25 | 92.5 | 65 | 290 | 105 | 925 | 145 | 2 900 | 185 | 9 250 | 225 | 29 000 | 265 | 92 500 |
| 26 | 95 | 66 | 300 | 106 | 950 | 146 | 3 000 | 186 | 9 500 | 226 | 30 000 | 266 | 95 000 |
| 27 | 97 | 67 | 307 | 107 | 975 | 147 | 3 075 | 187 | 9 750 | 227 | 30 750 | 267 | 97 500 |
| 28 | 100 | 68 | 315 | 108 | 1 000 | 148 | 3 150 | 188 | 10 000 | 228 | 31 500 | 268 | 100 000 |
| 29 | 103 | 69 | 325 | 109 | 1 030 | 149 | 3 250 | 189 | 10 300 | 229 | 32 500 | 269 | 103 000 |
| 30 | 106 | 70 | 335 | 110 | 1 060 | 150 | 3 350 | 190 | 10 600 | 230 | 33 500 | 270 | 106 000 |
| 31 | 109 | 71 | 345 | 111 | 1 090 | 151 | 3 450 | 191 | 10 900 | 231 | 34 500 | 271 | 109 000 |
| 32 | 112 | 72 | 355 | 112 | 1 120 | 152 | 3 550 | 192 | 11 200 | 232 | 35 500 | 272 | 112 000 |
| 33 | 115 | 73 | 365 | 113 | 1 150 | 153 | 3 650 | 193 | 11 500 | 233 | 36 500 | 273 | 115 000 |
| 34 | 118 | 74 | 375 | 114 | 1 180 | 154 | 3 750 | 194 | 11 800 | 234 | 37 500 | 274 | 118 000 |
| 35 | 121 | 75 | 387 | 115 | 1 215 | 155 | 3 875 | 195 | 12 150 | 235 | 38 750 | 275 | 121 000 |
| 36 | 125 | 76 | 400 | 116 | 1 250 | 156 | 4 000 | 196 | 12 500 | 236 | 40 000 | 276 | 125 000 |
| 37 | 128 | 77 | 412 | 117 | 1 285 | 157 | 4 125 | 197 | 12 850 | 237 | 41 250 | 277 | 128 000 |
| 38 | 132 | 78 | 425 | 118 | 1 320 | 158 | 4 250 | 198 | 13 200 | 238 | 42 500 | 278 | 132 500 |
| 39 | 136 | 79 | 437 | 119 | 1 360 | 159 | 4 375 | 199 | 13 600 | 239 | 43 750 | 279 | 136 000 |

Appendix 2: LOAD INDEX; The Load Index (LI) is a numerical code associated with the maximum load a tyre can carry at the speed indicated by its Speed Symbol under service conditions specified by the tyre manufacturer.

## Appendix 3

## European Tyre Manufacturers Speed Markings for Radial Ply and Diagonal (Cross) Ply Construction Tyres

| Old speed marking | | New speed symbol | Maximum speed mph | kmh$^{-1}$ |
|---|---|---|---|---|
| SR | radial reinforced M + S (15" & below) | P | 93 | 150 |
| SR | radial reinforced M + S (16" & above) | Q | 99 | 160 |
| SR | radial M + S | Q | 99 | 160 |
| SR | radial reinforced | R | 106 | 170 |
| SR | radial | S | 113 | 180 |
| HR | radial M + S | T | 118 | 190 |
| HR | radial | H | 130 | 210 |
| VR | radial | V | 150 | 240 |
| ZR | radial | ZR | over 150 | over 240 |
| - | diagonal (16" & above) | L | 75 | 120 |
| V | diagonal | V | over 130 | over 210 |

## Appendix 4

### "E" Numbers

| | | | |
|---|---|---|---|
| E1 | Federal Republic of Germany | E12 | Austria |
| E2 | France | E13 | Luxemborg |
| E3 | Italy | E14 | Switzerland |
| E4 | Netherlands | E15 | German Democratic Republic |
| E5 | Sweden | E16 | Norway |
| E6 | Belgium | E17 | Finland |
| E7 | Hungary | E18 | Denmark |
| E8 | Czechoslovakia | E19 | Romania |
| E9 | Spain | E20 | Poland |
| E10 | Yugoslavia | E21 | Portugal |
| E11 | United Kingdom | E22 | USSR |

# CURRENT PROGRESS OF SYNTHETIC RUBBERS FOR AUTOMOTIVE APPLICATIONS

T. Ohishi

Zeon Chemicals Europe Ltd
Sully
South Glamorgan, CF6 2YU
United Kingdom

## 1 INTRODUCTION

The gross production of Japanese automotive cars in 1988 was 12.70 million units which is 26.1% of world production.[1] This figure can be attributed to cars having high performance, high reliability and low cost, and was achieved over a long period through hard work and perseverance.

The Japanese rubber industry has grown with the development of the automotive industry and has helped to support the Japanese economy. Japanese rubber consumption was 1 410 million tonnes in 1989,[2] of this 78% was used in the automotive industry (73% for tyres and tubes, 5% for industrial goods, figure 1). The split of synthetic rubbers in new cars is 75% for tyres and tubes, and 25% for non-tyre applications. Figure 2 shows the classification of synthetic rubber which consists of general purpose rubber mainly used for tyres, speciality rubber used for other purposes and thermoelastomer.

An automotive car has over 10 000 parts, about 500 parts being rubber. The rubber content is 3 or 4%, but is indispensible because of its extremely low Young's modulus of elasticity and resistance to large deformation. Oil resistant rubbers have been used for functional and safety purposes in and around automotive engines; they are a key technological area, becoming increasingly important for the design and development of automotive cars. Tyres and tubes are basic functional parts of automotive cars and their improvement is a continuing theme.

Figure 3 shows changing social circumstances and the flow of related regulations introduced since the 1970s. As a result the increased requirements on rubber properties has pushed rubber technology forward. All these changing circumstances have given rise to the use of small cars with wedge-shaped profiles resulting in higher engine component temperatures, sour gasoline and problems caused by gasoline denaturing existing rubbers. The use of new fuels and long life lubricating oils has also resulted in the deterioration and fatigue of contacted rubbers. On the other hand, the demands of economy and safety have increased research into reducing the low rolling resistance and increasing high wet skid resistance of automotive tyres; these are conflicting properties.

In this paper, a few examples of the new Japanese technology of synthetic rubber and its applications during the past ten years are discussed.

## 2 OIL RESISTANT RUBBER MATERIALS

### The Changing Social Circumstances and Needs for Oil Resistant Rubber Materials

Figure 4 summarises the changing circumstances of the automotive industry[3,4] and the effects on oil resistant rubber materials.

The exhaust gas regulations and the need for fuel economy have resulted in the introduction of electronic fuel injection control systems. This controls the amount of fuel supplied to the engine and is proportional to its speed; the surplus fuel being returned to the fuel tank. These regulations have also meant increased temperatures in the engine compartment causing souring of gasoline (by formation of radicals and oxidation of gasoline recyled with air). So the existing oil resistant rubbers have not been compatible with the fuel hose and the carburettor and fuel pump diaphrams etc. New lubricant oils,that is engine oil, power steering oil and gear oil,have also been developed to increase their efficiency and periods between changes. The rubber parts in the oil delivery system require the highest resistance to lubricating oil additives and oxidised (sour) oil which attack the rubber aggressively at high temperatures.

For noise regulation and reduced weight of cars, the timing belt which transfers the rotary driving force of the engine to the camshaft to operate the inlet and exhaust valves of the engine has changed from metal to polychloroprene (CR), but the increasing temperature in the engine compartment has necessitated improved durability of the timing belt.

## Example of Developed Technology: Highly Saturated Nitrile Elastomer (HSN)

The classification system of SAE J200 is intended for use with rubber products in automotive applications, figure 5, for example oil resistant rubbers. Acrylonitrile-butadiene co-polymer (NBR) is a typical oil resistant rubber having good balanced properties. Highly saturated nitrile elastomer (HSN) was designed by Nippon Zeon with partial hydrogenation of the double bonds in the butadiene units leaving some of the double bond units as cure sites, but without destroying the nitrile groups in the NBR in order to improve the heat resistance and durability.

The key was to find and develop a selective hydrogenation catalyst which did not hydrogenate the nitrile group or solvent, figure 6. The breakthrough was discovering that a large pore diameter catalyst carrier was more active than one with small diameter pores; the reverse of low molecular mass hydrogenation reactions.[5] Polymer properties such as heat and ozone resistance are affected by the degree of hydrogenation, changing drastically at over 80% hydrogenation. Increasing the degree of hydrogenation of NBR increases stretching crystallisability, figure 7. This suggests that very accurate control is needed in the hydrogenation reaction; hence the technology was developed.

The research and development of HSN was started in 1975, the pilot plant stage being reached in 1980, then co-developed with car producers and commercialised by Nippon Zeon in 1984. In 1990 Zeon Chemicals USA also began to produce HSN in Texas.

## Automotive Applications of HSN

The grades of commercialised HSN (Zetpol™) are shown in table 1. The heat resistance of Zetpol™ compared with NBR, CR and ACM are shown as the service temperature and life in figure 8.[6] Figure 9 shows that HSN has a much higher tensile strength than the other oil resistant elastomers.[7]

Figure 10 shows the key properties of three rubbers, using radar charts, each of the eight arms represents a property regarded as significant for fuel resistant rubbers. The further the location of the points on the arm from the centre of the circle, the better are the properties. It can be seen that HSN is superior to NBR and ECO in heat resistance, ozone resistance and sour gasoline resistance, that is closest to a circle. This clearly indicates that it has the best balance of properties for fuel applications.[8]

For improved power steering hose which is exposed throughout its life to lubricating oil, the cover and inner tube have been changed from CR and NBR to CSM and HSN respectively. By improving the heat resistance by 20 C (as measured by the impluse test[9] figure 11) the life of the hose was increased by a factor of three at 140 C.

Twelve properties of rubber regarded as significant for timing belts are illustrated in figure 12 and figures 13, 14 &15 illustrate the properties of three elastomers. With the requirement of higher operating temperatures for timing belts, it can be seen that HSN has a good balance of properties with its stable dynamic complex modulus over a wide temperature range, promising low temperature properties, good heat and ozone resistance, lower permanent set, slower flex crack initiation after ageing and good oil resistance. HSN timing belts have already been used in Japan and Europe because of their longer life.

To summarise, the scope of HSN in automotive applications continues to develop widely in the Japanese and European markets, figure16. In some of the applications HSN has replaced conventional rubber, other applications are completely new.

## 3 RUBBER MATERIALS FOR TYRES

The second oil crisis in 1978 caused fuel consumption regulations in Japan, this increased the popularity of radial tyres. New requirements for low fuel consumption have brought the need for low rolling resistance coupled with high wet-skid resistance for passenger tyre materials. In the last ten years, requirements for tyres have become more and more segmented and specialised including high speed, comfort and all season tyres etc. Tyre materials to cope with these requirements are now available,[10] figure 17.

### <u>High Vinyl Polybutadiene (VBR)</u>[11]

An important requirement of tyres is a good balance between fuel economy and safety. This is achieved through improved low rolling resistance, which consumes 14.4% of the car's energy[12] (figure 18) and high wet-skid resistance for the tyre tread which accounts for half of the rolling resistance of radial tyres,[13] figure 19.

The tyre tread generates inner heat from repeated deformation during the rolling cycle, figure 20. Low rolling resistance decreases the amount of kinetic energy converted to heat energy. Hence material with low rolling resistance should have a low tan $\delta$. While high wet-skid resistance is synonymous with high friction resistance when the car is driven on a wet road, more kinetic energy is dispersed as heat energy during skidding, figure 21. Material with high wet-skid resistance should have a high tan $\delta$. It is believed that these critical properties of passenger tyre materials are closely related to the glass transition temperature $T_g$ of the rubber (figure 22) and that low rolling resistance and high wet-skid resistance are contradictory to each other, so there is a trade-off relationship, figure 23.

According to careful studies on the structure-property relationship of polybutadiene rubbers with various micro- and macro-molecular structure, it has been found that $T_g$ is not the only decisive factor. Polybutadiene rubbers with high vinyl contents exhibit both the low heat build-up and high wet-skid resistance. VBR with 70% vinyl content has the best balance for passenger tyres (figure24) and was commercialised by Nippon Zeon in 1980.

The characteristics of VBR/CBR and VBR/IR blends have been studied. The relationship between the rebound resilience at room temperature (23 C) and the wet-skid resistance of the blend series compared with the conventional SBR/CBR blends is shown in figure 25. SBR/CBR series show a linear relationship with a negative slope, but the VBR/CBR and VBR/IR blend series show relationships with a convex upward curve. Consequently, VBR/CBR and VBR/IR blends show a good balance of these two properties shifted in a desirable direction (high rebound and high wet-skid resistance) for tyre application.

## Chemically Modified Solution Polymerised Rubber[15]

It was realised during the development of VBR that the improvement of polymer-filler interaction should create better properties for new type materials. A technique was developed by Nippon Zeon whereby solution polymers were chemically modified by adding polar compounds such as 4,4'-bis(diethylamino)-benzophenone (EAB) to the molecular chain ends using a lithium catalyst system without changing the main chain structure, figure 26. From absorption experiments, *n*-butyl-EAB, a model compound with a modified chain end, showed a strong interaction (chemisorption) with carbon black surfaces.

Chemically modified rubber enhances the dispersion stability of the carbon black in vulcanites, and hence suppresses the transient-structure formation, or network formation between carbon black particles. The reduction in the transient-structure decreases the dynamic shear modulus at low strain amplitude, and hence decreases the loss of modulus to give an increase in rebound resilience,[16] figure 27.

This chemical modification has been applied to both high vinyl polybutadiene (Nipol BR 1245) and solution polymerised SBR (Nipol SBR SN 100's). Figure 28 shows that the chemically modified SBR has not only high resilience (low rolling resistance) and high wet-skid resistance for low fuel consumption tyres, but also reduced hardness at -10 C. The reduced hardness at -10 C means a smaller increase in hardness at low temperatures, giving high skid resistance on ice and snow. Therefore, chemically modified SBR is suitable for all season tyre tread with low rolling resistance and high skid resistance on wet, snow or ice covered surfaces. Mass production of chemically modified SBR was started by Nippon Zeon in 1986, and has been used by automobile manufacturers where specialised tyres are needed.

## 4 SUMMARY

A significant portion of the Japanese Automotive Industry growth can be attributed to the production of high performance, high reliability cars. For this growth, a three way joint development by car, parts and raw material producers has been of some importance. When a new polymer is developed the research can begin about ten years beforehand, hence it is very important to anticipate future needs and develop responses jointly. The car industry should fulfil customers needs of variety and individuality as much as possible. Material technology is based on the development of higher performance and better suited materials which support the development of new parts and machines which are then incorporated into new car designs,[17] figure 29.

## REFERENCES

1. Toyota Automotive Co. Ltd, 'Toyota 1989 - Outline of Automotive Industry', Aichi Prefecture, March 1989, 11.
2. Statistical Report, The Society of Japanese Rubber Industry.
3. A. Yoshioka, *Polymer Preprints*, Japan, 1988, **37**, 13
4. T. Ohishi, *Rubber Industry*, 1987, **23**(4), 20.
5. Y. Kubo, *Journal of the Society of Rubber Industry*, Japan, 1986, **59**, 442.
6. Nippon Zeon Co. Ltd, "Zetpol" catalogue.
7. K. Hashimoto, N. Watanabe and M. Oyama, 127th ACS Rubber Division Meeting, Los Angeles, 24th April 1985.
8. K. Hashimoto, M. Oyama, M. Watanabe and Y. Todani, 128th ACS Rubber Division Meeting, Cleveland Ohio, 1 - 4th October 1985.
9. Y. Tsuzuki, *Journal of the Society of Rubber Industry*, Japan, 1989, **62**, 650.
10. T. Ohishi, 36th Summer University of High Polymer, The Society of Polymer Science, Japan, Miyagi, Zao 13 - 16th July, 1988, 49.
11. A. Ueda, S. Akita, H. Watanabe, K. Komuro and A. Yoshioka, *Polymer Preprints*, Japan, 1987, **36**, 79.
12. Y. Kimbara and A. Sigiura, *Journal of the Society of Automobile Engineering*, Japan, 1978, **32**, 379.

13. N. Yoshimura, M. Okumura and Y. Yamagishi, 122nd ACS Rubber Division Meeting, Chicago, 5 - 7th October 1982.
14. E.F. Engel, *Gummi Asbest Kunststoff*, 1972, **26**, 362.
15. A. Yoshioka, A. Weda, U. Watanabe and N. Nagata, *The Chemical Society of Japan*, 1990, **4**, 341.
16. N. Nagata, 131st ACS Rubber Division Meeting, Montreal, Quebec, Canada, 26 - 29th May 1987.
17. Y. Tsuzuki, *Journal of the Society of Rubber Industry*, 1989, **62**, 650.

## Table 1. Commercial grade of Zetpol

| Grade | Bound ACN (%) | Iodine Value (g/100g) | Mooney Viscosity ($ML_{1+4}$100°C/212°F) |
|---|---|---|---|
| Zetpol 1020 | 44 | 25 | 78 |
| Zetpol 1010 | 44 | 10 | 85 |
| Zetpol 2030L | 36 | 58 | 57.5 |
| Zetpol 2020 | 36 | 28 | 78 |
| Zetpol 2020L | 36 | 28 | 57.5 |
| Zetpol 2010 | 36 | 11 | 85 |
| Zetpol 2010L | 36 | 11 | 57.5 |
| Zetpol 2000 | 36 | 4 | 85 |
| Zetpol 2000L | 36 | 4 | 65 |

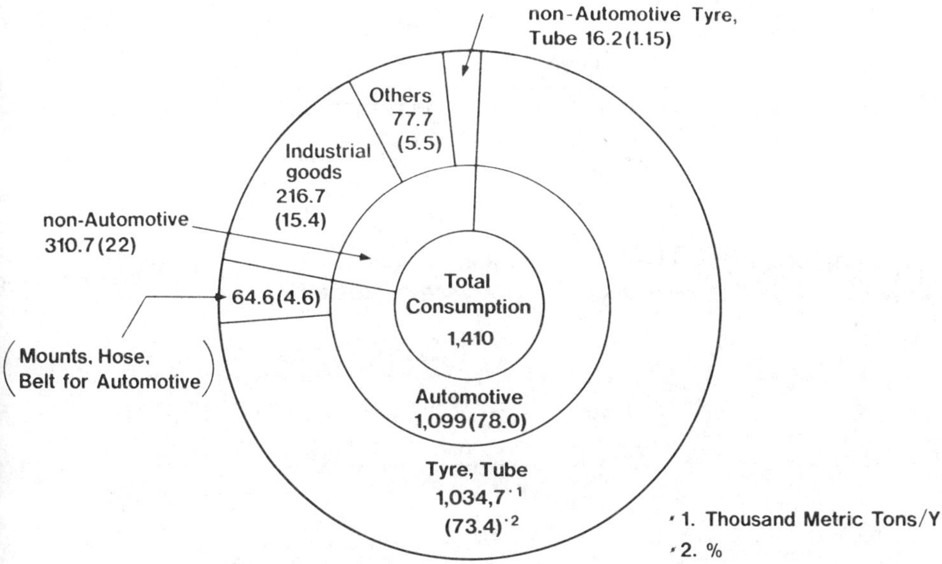


Figure 1 **Breakdown of New Rubber Consumption in Japan (1989)**

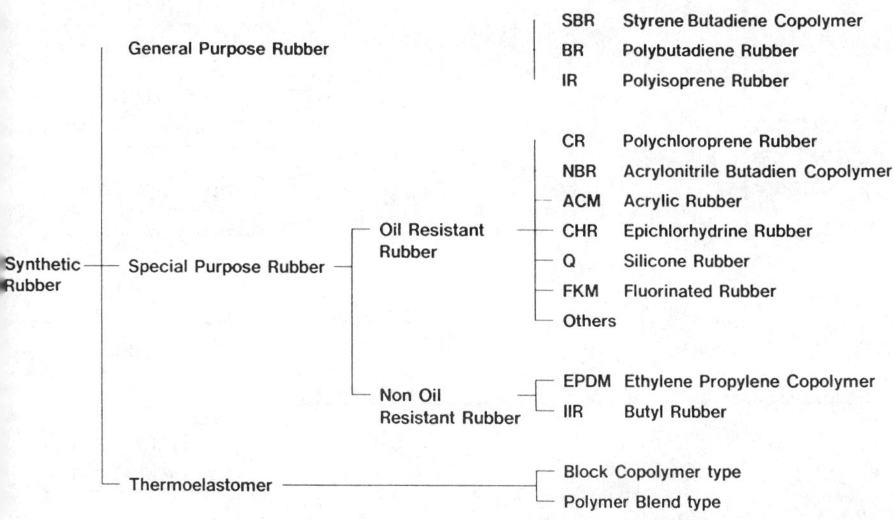


Figure 2 **The Classification of Synthetic Rubber**

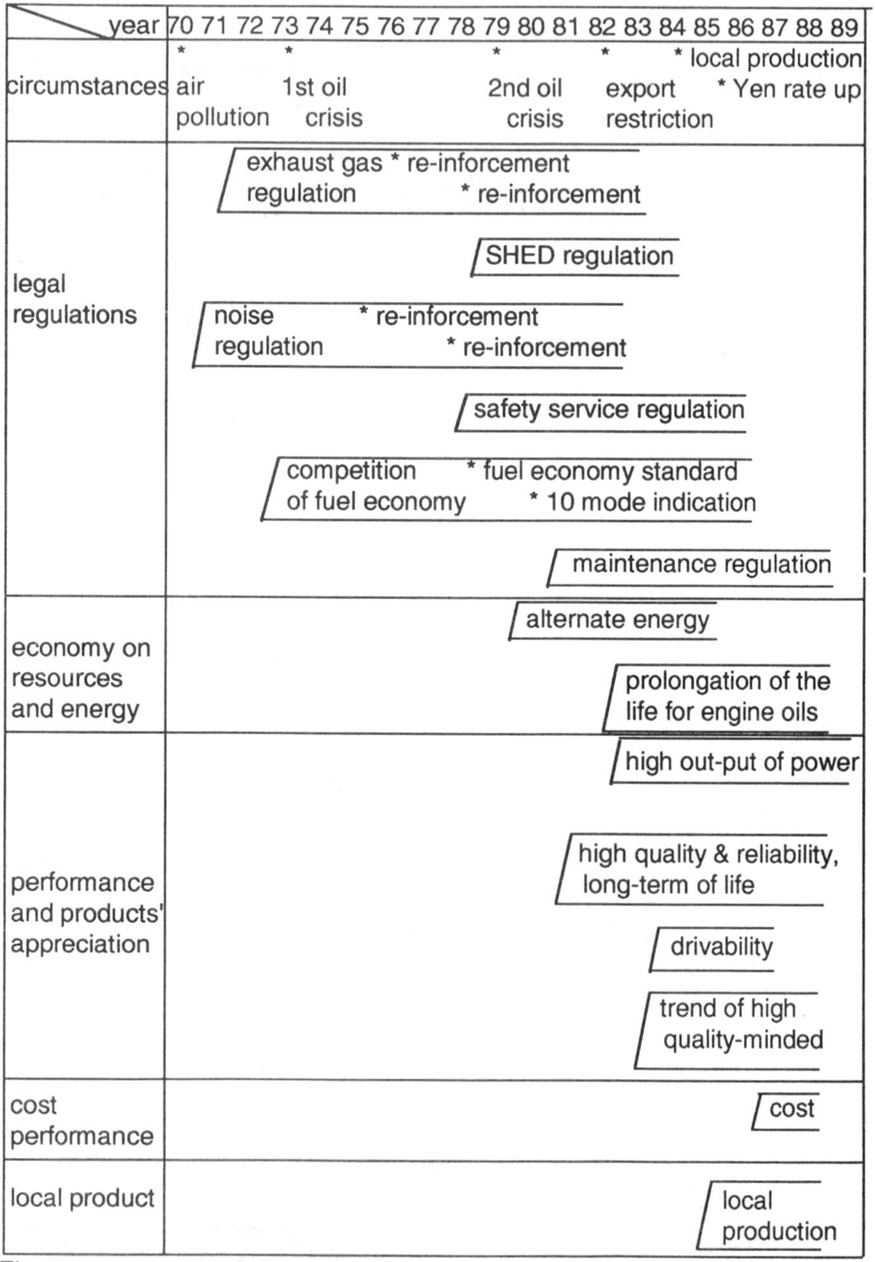


Figure 3 Current trends of automotive cars and requirements for elastomers

| counter measures | requirements for elastomers |
|---|---|
| improvement of engines<br>EGR, catalyst converters<br>lead free gasoline (high aromatic fuel) | heat resistance<br><br>solvent crack resistance |
| canister system<br>prevention of leakage of gasoline | gasoline permeability (low) |
| design of noise insulation | low dynamic stiffness ratio<br>high damping property |
| impact insulating bumpers<br>impact insulating steering | wet skid resistance |
| lightened weight, design tyres for fuel economy, reduction of friction (low viscosity engine oils) | heat resistance (small-sized engine room), low rolling resistance (tyres) oil resistance |
| improve materials<br>adoption of new polymers | heat, additive, ozone, sour oil and fatigue resistance |
| ethanol-methanol mixed gasoline | alcohol mixed gasoline resistance |
| new kind of additives | duration against high performance engine oils, sour oil resistance, metal sludge resistance |
| PGM-FI, turbo, DOHC, 4VLV, V6<br>super charger | sour gasoline & heat resistance |
| improvement of materials, adoption of new polymers, quality control, long life fluid (SF oil, LLC) | heat, ozone, fatigue & lubricant additive resistance |
| 4WD, 4WALB, 4WS | (development of specialty devices) |
| reduction of vibrating noise, improvement of exterior materials, designs | low dynamic stiffness ratio, high damping property, non-carbon compounds, heat resistance |
| alternative materials, VA (value analysis) | cost-down, heat resistance<br>(total cost-down) |
| local purchasing | internationalisation of rubber technology |

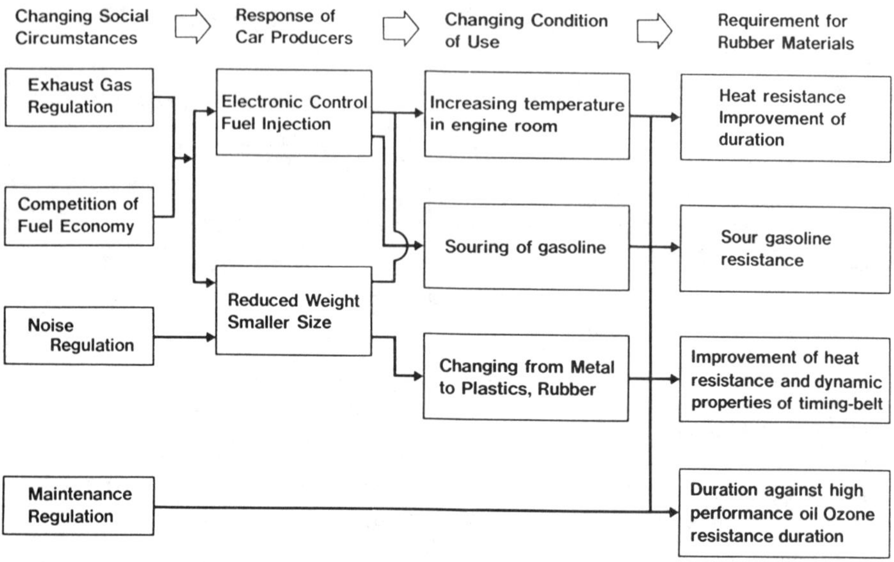


Figure 4 **The Changing Circumstances of The Automotive Industry and Effect on Oil Resistant Rubber Materials**

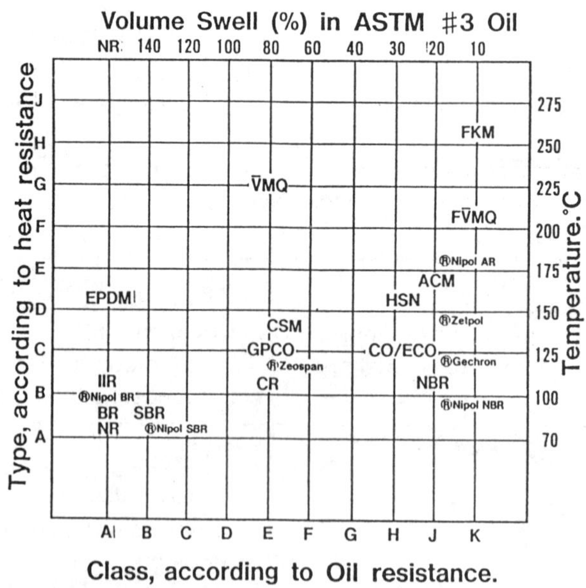


Figure 5 **Classification of HSN by SAE J200**

**1. COPOLYMERIZATION OF ETHYLENE WITH ACRYLONITRILE**

$$\underset{E}{CH_2{=}CH_2} + \underset{AN}{CH_2{=}\underset{C\equiv N}{\underset{|}{CH}}} \rightarrow -CH_2-CH_2-CH_2-\underset{C\equiv N}{\underset{|}{CH_2}}-$$

**Very Dififcult Copolymerization ← Reactivities : AN ≫ E**

**2. SELECTIVE HYDROGENATION OF NITRILE RUBBER**

$$-(CH_2-CH{=}CH-CH_2-)x-(CH_2-\underset{C\equiv N}{\underset{|}{CH}}-)y-$$

$$H_2 \quad \Downarrow \quad \text{Catalyst}$$

$$-(CH_2-CH_2-CH_2-CH_2-)x-(CH_2-\underset{C\equiv N}{\underset{|}{CH}}-)y-$$

**• Having to develop the highly active and selective hydrogenation reaction catalyst**

Figure 6 **SELECTION OF PRODUCTION METHOD**

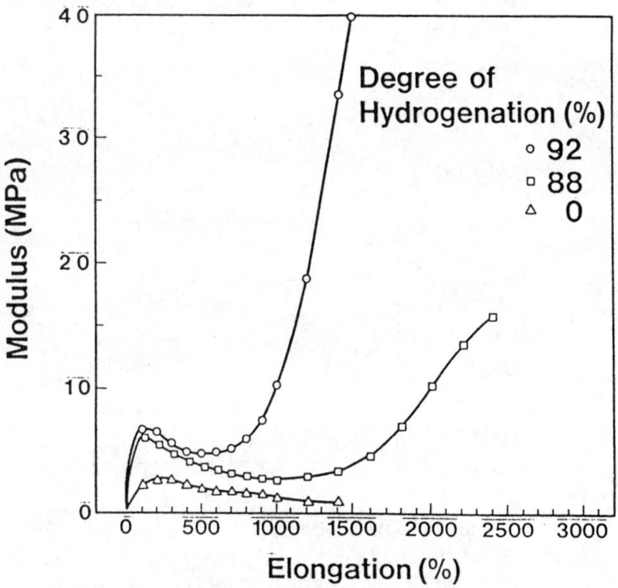


Figure 7 **Stress Strain Curves**

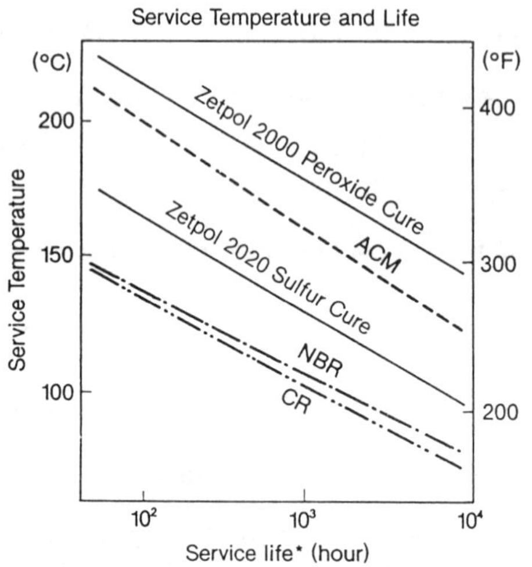


* Service life assumed the time to minus 80% elongation change on ageing test in air.

Figure 8

## Heat Resistance

Grades such as Zetpol 1020 and 2020 with iodine values of 20~30 when sulfur cured, can be continuously used for 1,000 hrs at 125°C (257°F). On the other hand Zetpol 2000 cured with peroxide can be continuously used for 1,000 hrs at 160°C (320°F).

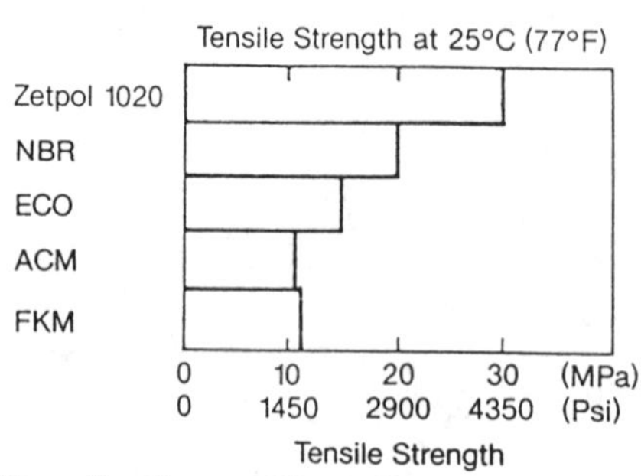


Figure 9

## Tensile Properties

Zetpol has higher tensile strength and wear resistance than NBR, to say nothing of the conventional heat, and oil resistant elastomers.

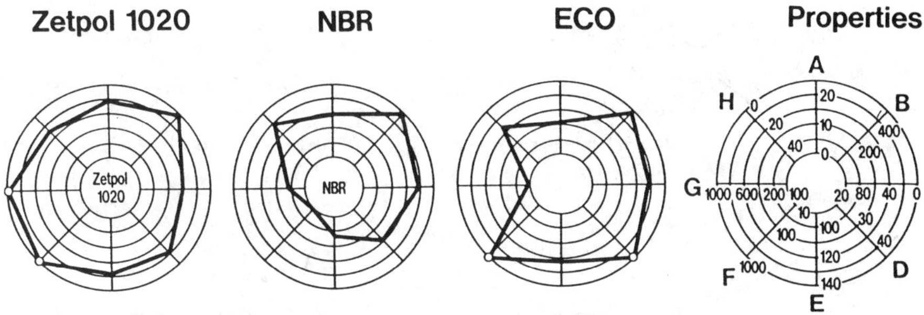


A : Tensile strength (MPa)
B : Elongation (%)
C : Compression set 70 hrs×120°C
D : Brittleness temperature (°C)
E : Service temperature at 100% elongation after ageing for 1,000 hours
F : Time to crack initiation by ozone 50 pphm 20%
G : Time at 5 MPa tensile strength in oxidized fuel B with 1% LPO at 60°C
H : Volume change in fuel C 48hrs×40°C

Figure 10 **Properties of Fuel Resistant Elastomers**

| | Materials | | | Endurance life of impulse test (1 2 3 4) |
|---|---|---|---|---|
| | Cover | Inner tube | Reinfored fiber | |
| Conventional Hose | CR | NBR | Nylon 66 | 120°C (bar to 3); 140°C (bar to 1) |
| Improved Hose | CSM | HNBR | Nylon 66 | 120°C (arrow beyond 4); 140°C (bar to 3) |

Tsuzuki Y : Journal of the Society of Rubber Industry, Japan 650 62, 1989

Figure 11 **The Heat Resistance and Endurance Life of Power Steering Hose**

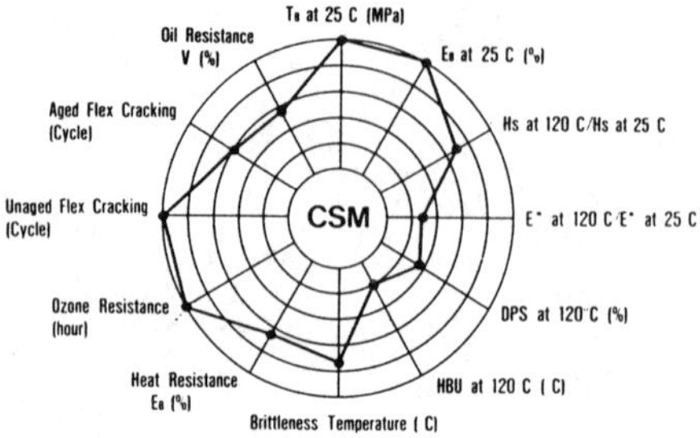


**Key Properties of CSM for Timing Belt**

Figure 12

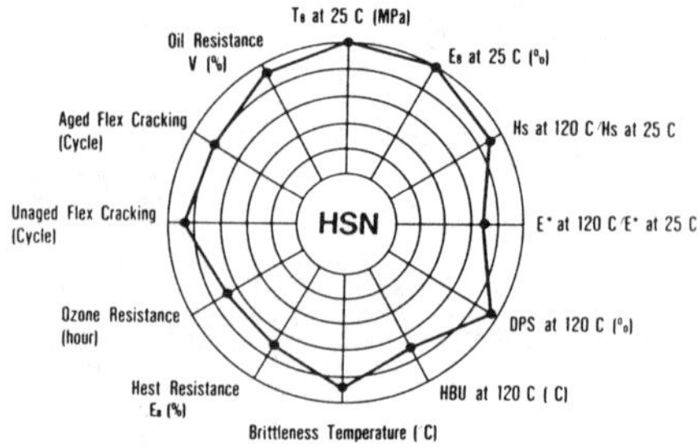


**Key Properties of HSN for Timing Belt**

Figure 13

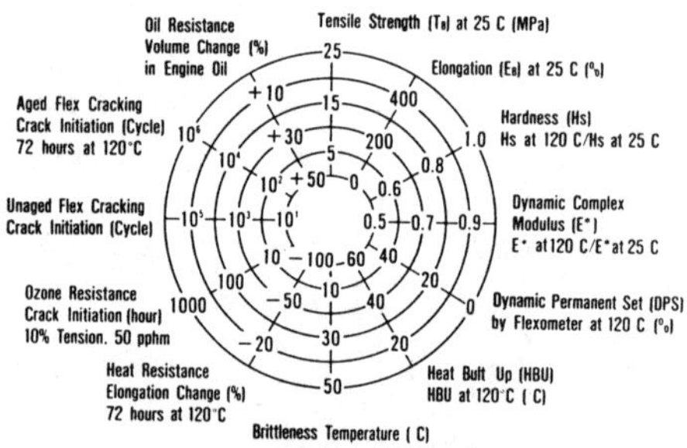


**Key Properties for Timing Belt**

Figure 14

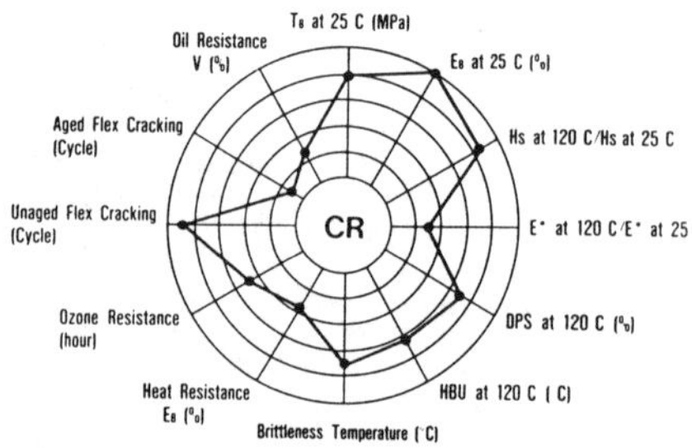


**Key Properties of CR for Timing Belt**

Figure 15

| Applications | Current Materials |
|---|---|
| **BELTS** | |
| Timing Belt | CR<br>CSM |
| Poly-V-Belt | CR |
| CVT Belt | – |
| Balancer Belt | – |
| **FUEL SYSTEM** | |
| Fuel Hose | NBR<br>PB, FKM |
| In-Tank Insulator | FKM<br>PB |
| Cab. Insulator | NBR |
| **DIAPHRAGM** | |
| Fuel Pump DP | NBR |
| Distributor DP | NBR |

| Applications | Current Materials |
|---|---|
| ◦ **OIL TUBE and HOSE** | |
| Power Steering Hose | NBR |
| Oil Cooler Hose | NBR<br>ACM, ECO |
| Freon Hose | NBR |
| Emission Cont. Tube | ECO |
| ◦ **SEAL, O-RINGS and PACKINGS** | |
| Freon Comp. Seal | FKM |
| Water Pump Seal | FKM |
| Bearing Seal | FKM |
| Hub Seal | X-NBR |
| P.S Seal | NBR |
| Valve Stem Seal | FKM |
| Various O-Rings | NBR |
| for Freon Hose | ECO |

| Applications | Current Materials |
|---|---|
| ◦ **GASKETS** | |
| Metal Head GSK | FKM |
| ◦ **CHAIN TENSIONER** | |
| Chain Tensioner | NBR<br>Tefron |
| ◦ **ANTI-VIBRATION and DAMPING** | |
| High Hs Bushing | – |
| Engine Mount | – |
| Gear Dumper | – |
| ◦ **CLUTCHING FACING** | |
| Clutching Facing | XNBR |
| ◦ **Muffler Hunger** | – |
| ◦ **CYLINDER LINER** | Q |

Figure 16 **Development of Automotive Application for Hydrogenated NBR, Zetpol in Japanese and European Markets**

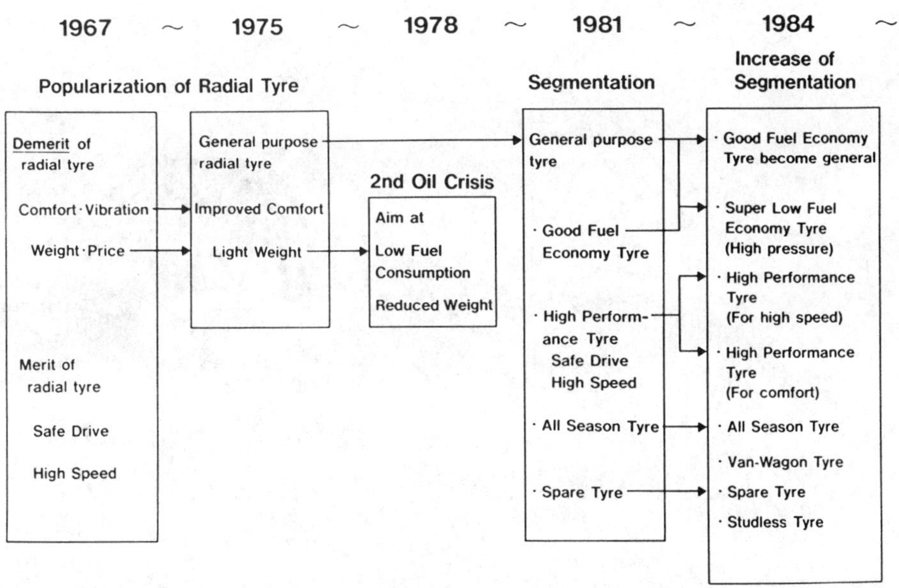


Figure 17 **Changing Needs for Passenger Tyres**

**1. Recent reqirement for tyre performance**

☆ **Low Rolling Resistance**
→ **Economy**

☆ **High Wet Skid Resistance**
→ **Safety**

☆ **High speed Durability**

☆ **Driving stability**

☆ **Ride comfort**

**2. Energy Consumption of Car by 10 Mode Driving**

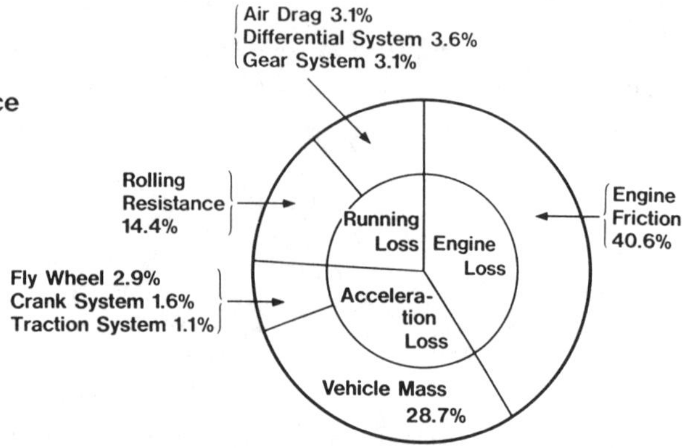


Figure 18 **Necessity of Low Fuel Consumption Tyre**

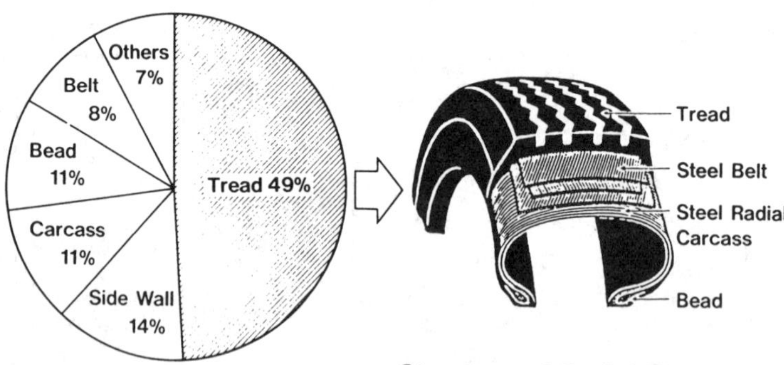


**Structure of Radial Tyre**

Figure 19

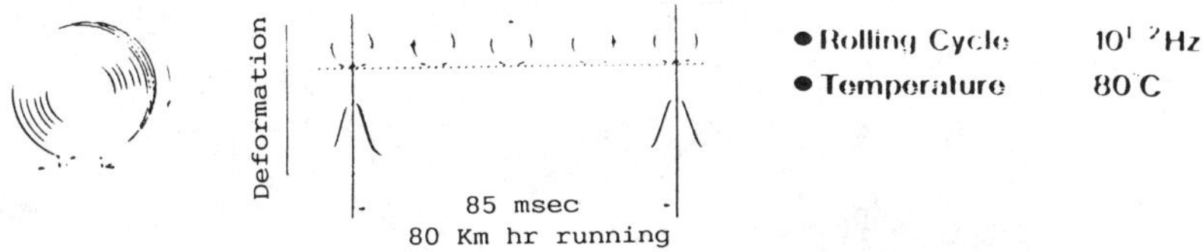


Figure 20 **Deformation of a tyre during rolling (Rolling Resistance)**

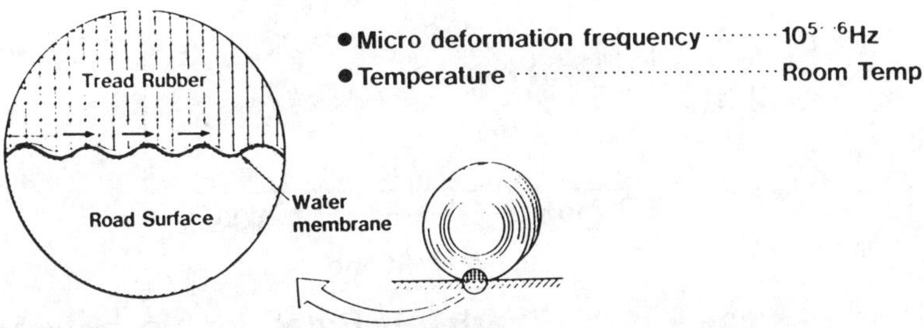


Figure 21 **Deformation of a tyre tread surface during sliding (Wet Skid Resistance)**

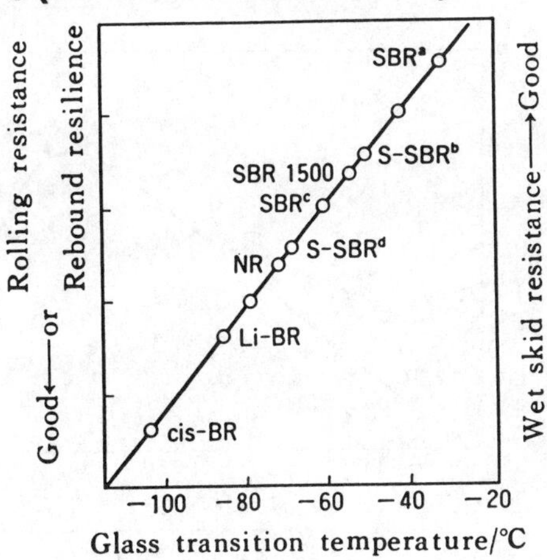


Relation between $T_g$ of typical rubbers and tire tread rubber properties[6)]

Figure 22 SBR[a] : 40% styrene, S-SBR[b] : 25% styrene, 25% vinyl, S-SBR[c] : 25% styrene, 8% vinyl, S-SBR[d] : 19% styrene, 8% vinyl, S : solution polymerized

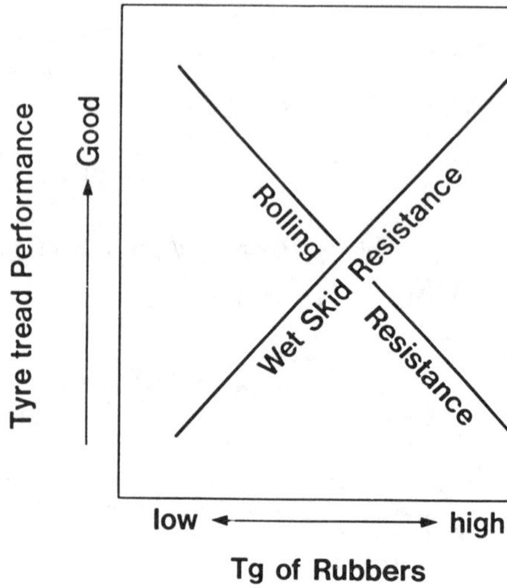


Figure 23 **Trade-off Relationship between Rolling Resistance and Wet Skid Resistance for Tyre Tread Rubbers.**

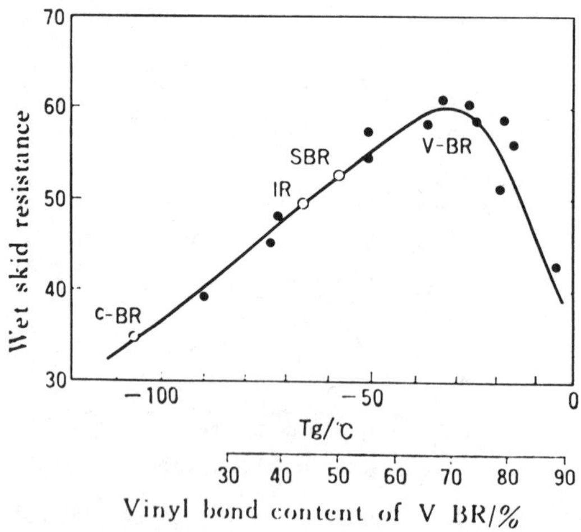


Figure 24

Relation between wet skid resistance and $T_g$ on several types of rubbers and various V BR

c BR : High *cis* 1,4-polybutadiene rubber, SBR : Styrene butadiene rubber, IR : High *cis* 1,4-polyisoprene rubber, V-BR : Vinyl polybutadiene rubber

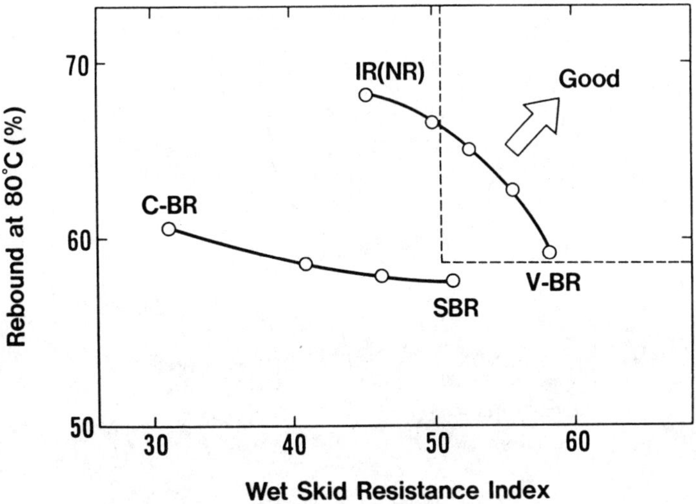


Figure 25 **Relationship between W.S.R and Rebound to HVB/NR Blend Series**

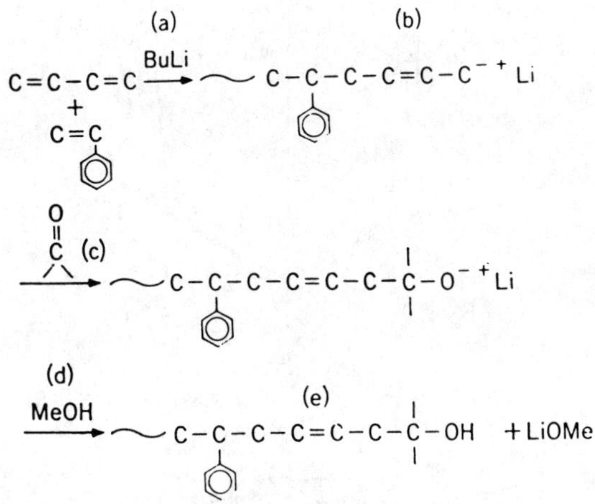


Chemical modification reaction scheme

Figure 26 a : Catalyst, b : Polymerization degree, 2000∼5000, c : Modifier, d : Stopper, e : Chemically modified rubber on chain end

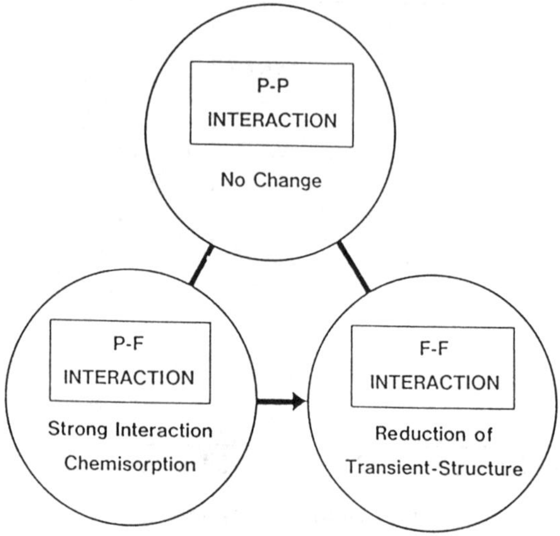


Figure 27 **Interactions in the Black-Loaded Vulcanizate and Change in the Interactions of the Modified Rubber**

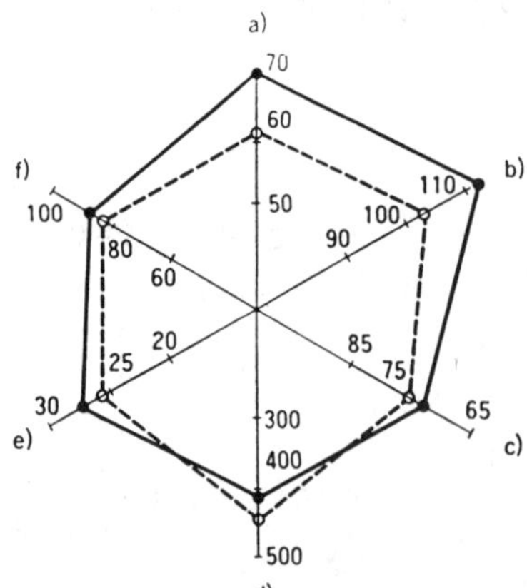


Figure 28

Typical physical properties of S-SBR on the tread of all-season tire

● : Modified, ○ : Unmodified

a ) Rebound resilience at 60℃
b ) Wet skid resistance index (SBR 1502=100)
c ) Hardness at −10℃(JIS A)
d ) Ultimate elongation/%
e ) Tensile strength/MPa
f ) Abrasion resistance (SBR 1502=100)

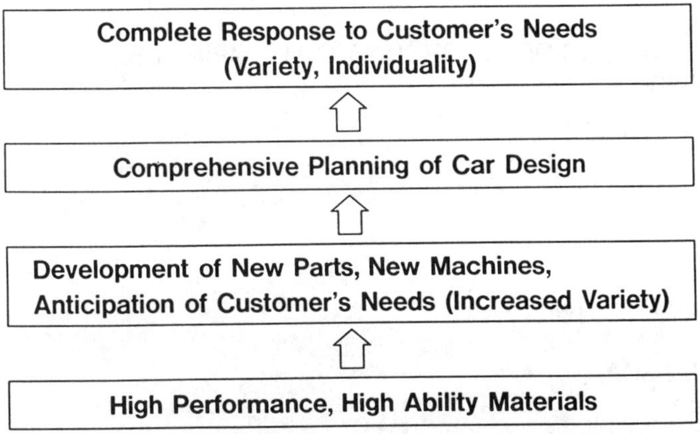


**Tsuzuki, Y : Journal of the Society of Rubber Industry, Japan 650 62, 1989**

Figure 29 **The Role of Materials Technology for Automotive**

# NOVEL APPLICATIONS OF CO-ORDINATION COMPLEXES: ANTI-CORROSIVES

B. Balmforth

ICI Colours and Fine Chemicals
Blackley
Manchester, M9 3DA
United Kingdom

## 1 INTRODUCTION

The automotive industry faces the on-going challenge of meeting the increasing demands of new technologies including, for example, more severe operating conditions, extended service life, more efficient usage of materials etc. As part of this challenge, rust prevention is no longer regarded as a marginal benefit, but is being viewed increasingly as essential, across many industries including the automotive industry, especially in situations where conditions are not favourable. These include areas of exposure to aggressive media or conditions during either construction or use of automobiles and components.

Iron and steel rust when in contact with air and water, and this becomes especially serious when the water contains salts or acids. Traditionally control of rust is achieved by incorporation of suitable rust inhibitors into the product composition. Typically these are based on traditional products such as phosphoric or tannic acids, sulphonates, amines etc. depending on the particular application. However, many situations arise where rust prevention afforded by these conventional inhibitors is barely adequate, or indeed wholly inadequate. This places heavy demands on the skills of the formulators, since simply increasing the level of rust inhibitors either does not give the desired effect, or leads to antagonism with other additives or exerts a detrimental effect on other properties of the formulation.

As part of a major initiative by ICI Speciality Chemicals into the rust preventative field, novel, patented rust preventatives have now been identified[1] and developed with industrial partners. These new products derive from ICI's expertise in co-ordination chemistry which has previously generated major businesses in dyestuffs, pigments, mining chemicals etc. This brings a new approach to rust prevention, and has already provided novel corrosion inhibition actives for the industry formulator to use in a wide range of product areas, as will be discussed in more detail later.

Previous work within ICI has demonstrated that metal surfaces can be modified via the careful selection of organic molecules which interact with metals, so called co-ordination chemistry, particularly those with surface recognition properties. This allows a "tailored" approach to this area of organic chemistry, to produce "designer molecules" to give anti-corrosive or other properties whilst ensuring compatibility with the desired medium.

Recent research and development work aimed at improving polymer adhesion to metals via surface recognition agents is described in the accompanying paper by J.R. Lawson. The novel rust preventatives described in this paper have resulted from this approach, and offer improved levels of rust prevention over conventional products across a range of different types of application. They offer a range of exciting new actives and concentrates for the skilled industrial formulator to utilise.

## 2 REVIEW OF EXPERIMENTAL DATA

By functionalising the various groups on the novel co-ordination compounds, a range of versatile actives can be produced, designer made for the individual application. Characteristics of these products, together with data on their performance in standard tests etc. are presented for a range of applications relevant to this conference. These data have been obtained mainly in-house by ICI, but independent validations are also included where appropriate (see references).

Data are attached in tabulated form as Appendices 1 to 6, which exemplifies their use in:

a) rust converters,
b) anti-corrosive pigments,
c) lubricating greases,
d) mineral oil,
e) synthetic oils,
f) temporary protection.

## 3 DISCUSSION

Whilst traditional rust preventative products have been judged to be satisfactory for many years, today this is no longer the case. They are being seen as ineffective against increasingly harsh operating conditions, and many are coming under scrutiny because of the impact of environmental issues. Specific levels of rust prevention are becoming more difficult to meet, and are often only achieved at the expense of accommodating some detrimental effect on other properties. In short, times have been changing; performance requirements have been changing; but the basic chemistry of the existing actives for rust prevention has not!

The new rust preventative materials based on organic co-ordination chemistry described in this paper offer a new chemical approach to achieving improved levels of rust prevention and in so doing, overcome many of the practical limitations and concerns outlined earlier. All the products are EINECS and TSCA registered and have been commercialised by ICI Speciality Chemicals under the generic ATRUST™ trademark.

## 4 CONCLUSIONS

The new rust preventative materials, based on novel patented, organic-metal-complexing agents offer the automotive industry formulators valuable assistance in meeting the demands for improved rust prevention across a wide range of applications. They satisfy the key industry requirements of:

a) improved rust prevention (high performance/long life),
b) ease of formulation,
c) cost effectiveness,
d) safety (toxicity and environmentally).

On this basis the outlook for these novel rust preventatives looks very promising.

## REFERENCES

1 ICI portfolio of anti-corrosive (Atrust related) patents and applications.
2. J.D. Scantlebury and G.R. John et al, 'Metal complex agents as possible film forming anti-corrosives on mild steel', Proceedings of the International Symposium in Advances in Corrosion Protection by Organic Coatings, Cambridge, 1989.
3. Report of Paint Research Association, October 1989, (copies available ex ICI Speciality Chemicals).
4. Report of Paint Research Association, November 1988, (copies available ex ICI Speciality Chemicals).

5. N.A.C.E. Proposed Technical Committee report on Conversion Coatings.
6. G.R. John and B. Tury, 'Novel Rust Preventatives', N.L.G.I. 56th Annual Meeting, St. Louis, USA, October 1989.

## APPENDIX 1

RUST CONVERTERS[1,2,3,5]

Performance*

| Salt spray (ASTM B117) | 250 hrs | 500 hrs | 1000 hrs |
|---|---|---|---|
| typical tannic acid based | 1 | 0 - 1 | 0 - 1 |
| ICI ATRUST™ RC50H/10H† | 9 - 10 | 8 - 9 | 5 - 6† |

| Prohesion | | | |
|---|---|---|---|
| typical tannic acid based | 8 | 6 | 2 - 3 |
| ICI ATRUST™ RC50H/10H† | 10 | 8 - 9 | 5 - 6† |

| Storage stability (accelerated, 50 C) | 30 days | 60 days | 150 days |
|---|---|---|---|
| typical tannic acid based | OK | fail | fail |
| ICI ATRUST™ RC50H/10H† | OK | OK | OK |

| Overcoatability (with water based topcoats) | bleedthrough |
|---|---|
| typical tannic acid based | yes |
| ICI ATRUST™ RC50H/10H† | no |

* - rust ratings ASTM D610 (10 = no rust, 0 = all rust).
† - ICI ATRUST™ products are based on substituted salicylaldoximes.

## APPENDIX 2

## ANTICORROSIVE PIGMENT[1,4]

| Underfilm rusting (prohesion)* | 3% on D.F. | 6% on D.F. |
|---|---|---|
| blank | 7 | 7 |
| typical zinc phosphate | 6 | 4 |
| commercial specialty (K white 84) | 6 | 4 |
| ICI ATRUST™ 9600† | 3 | 2 |

| Rusty steel protection* | natural exposure (12 months) |
|---|---|
| commercial specialty (K white 84) | 6 |
| commercial specialty (shieldex) | 5 |
| optimised zinc phosphate | 4 |
| red lead | 1 - 2 |
| ICI ATRUST™ 9600* | 1 - 2 |

| Clean steel protection* | prohesion | natural exposure (12 month) |
|---|---|---|
| commercial specialty (K white 84) | 5 | 5 |
| commercial specialty (shieldex) | 6 | 5 |
| optimised zinc phosphate | 5 | 4 |
| ICI ATRUST™ 9600 | 3 | 2 |

* - all assessments on Champion scale (7 = worst, 0 = best)
† - ICI ATRUST™ product based on a salt of a substituted hydroxy carboxylic acid

## APPENDIX 3

## LUBRICATION GREASE[1,6]

S.K.F. EMCOR test (IP 220)*

| Lithium soap grease | water | 3% brine | pulp & paper water |
|---|---|---|---|
| base grease | 1 | 5 | 4 |
| petroleum sulphonate | 1 | 5 | 2 |
| synthetic sulphonate | 1 | 5 | 3 |
| ICI ATRUST™ 9013/9052† | 0 | 0 | 0 |
| **Calcium soap grease** | | | |
| base grease | 0 | 5 | - |
| fully form. commercial grease | 0 | 2 | - |
| ICI ATRUST™ 9100§ | 0 | 0 | - |
| **Organo clay grease** | | | |
| base grease | 5 | - | - |
| typical commercial D.S.S. | 3 | - | - |
| ICI ATRUST™ 9650¶ | 0 | - | - |
| **Specialty silica/silicone** | | | |
| base grease | 5 | - | - |
| synthetic sulphonate | 5 | - | - |
| typical commercial D.S.S. | 2 | - | - |
| ICI ATRUST™ product¥ | 0 | - | - |

N.B. Key physical properties of greases were unaffected.

* - results on scale 5 = worst, 0 = perfect.
† - ATRUST™ 9013/9052 are synergistic mixtures containing a substituted salicylaldoxime.
§ - ATRUST™ 9100 contains a co-ordinating polyester amine.
¶ - ATRUST™ 9650 contains a salt of an aromatic acid.
¥ - ATRUST™ product is a synergistic mixture containing a chelating active.

## APPENDIX 4

MINERAL OIL[1]

<u>ASTM D665 using 3% brine and 100% solvent neutral base</u>

| Additive | % protection | |
|---|---|---|
| | 25 ppm | 50ppm |
| typical petroleum sulphonate | 30 | 60 |
| typical synthetic sulphonate* | 60 | 70 |
| ICI ATRUST™ 9013† | 90 | 95 |

* - salt of dinonyl naphthalene sulphonic acid.
† - ICI ATRUST™ 9013 is a synergistic mixture containing a substituted salicylaldoxime.

## APPENDIX 5

### SYNTHETIC LUBRICATING OILS*

| ASTM D665 | water | 3% brine |
|---|---|---|
| **Water miscible** | | |
| blank | fail | fail |
| ICI ATRUST™ product I† | pass | pass |
| **Water immiscible** | | |
| blank | fail | fail |
| ICI ATRUST™ product II† | pass | pass |

| Typical dip-coat application | immersion test (days) | humidity test (days) |
|---|---|---|
| blank | <1 | 4 |
| commercial product A | 4 | <2 |
| commercial product B | 4 | 4 |
| ICI ATRUST™ product III† | >14 | 23 |

| Typical immersion application | immersion test (14 days, 60 C) |
|---|---|
| blank | fail |
| commercial product | fail |
| ICI ATRUST™ product IV† | pass |

* - typically polyalkylene glycols.
† - range of ICI ATRUST™ products based on co-ordination chemistry.

## APPENDIX 6

## TEMPORARY PROTECTION CLEAN C.R. STEEL[1]

| | salt spray ASTM B117 | condensing humidity (h) | natural exposure (weeks) |
|---|---|---|---|
| Typical dip-coat application | | | |
| commercial product | 20 h - 30% rust | 500 | 3 |
| ICI ATRUST™ 9502* | 200 h - 0% rust | 1000 | 12 |
| Slushing oil application | | | |
| synthetic sulphonates† | 20 h - 50% rust | - | - |
| ICI ATRUST™ 1004§ | 20 h - 0% rust | - | - |
| Car wax application | | | |
| a) Clean steel | | 24 h | |
| commercial product A | - | extensive rust | - |
| commercial + 1% ATRUST™¶ | - | no rust | - |
| b) Scribed car body panel | | 48 h | |
| commercial product A | - | rust in scribe | - |
| commercial product B | - | rust in scribe | - |
| formulation 1% ATRUST™¶ | - | no rust in scribe | - |

* - ICI ATRUST™ 9502 is a solvent formulation of a co-ordinating polyester amine.
† - salt of dinonyl naphthalene sulphonic acid.
§ - ICI ATRUST™ 1004 is a formulated co-ordinating amine ethoxylate.
¶ - ICI ATRUST™ product is a formulated salt of an organic carboxylic acid.

# NOVEL APPLICATIONS OF CO-ORDINATION COMPLEXES: ADHESIVE PROMOTERS

J.R. Lawson, P.A. Tasker and D. Stewart

ICI Colours and Fine Chemicals
Blackley
Manchester, M9 3DA
United Kingdom

## 1 INTRODUCTION

The problem of poor adhesion and durability when adhesive bonds are exposed to corrosive environments is well documented.[1] Many metal pretreatments have been developed in attempts to obviate these problems, including grit-blasting, oxidation and chemical etching processes. All these processes are designed to increase the surface area of the metal and thus enable better "keying" to the adhesive. Some chemical pretreatments have evolved, using silanes, titanates etc. as adhesion promoters for a variety of metals. These materials are generally difficult to apply, have limited stability and do not work efficiently over a range of metals and polymers. Inorganic materials such as chromates and plumbates are effective but suffer from the problem of extreme toxicity. Other approaches encourage the careful design of bonded joints in order to minimise stresses within the adhesive.[2]

ICI's strategy, developed within the Research Centre at Blackley, seeks to address the problem of poor adhesion/durability to metals by adopting a "designer adhesion promoter" approach, whereby adhesion promoters are "tailor-made" for a specific metal-polymer system. Powerful ligating groups specific to a particular metal are firstly identified by testing a large number of candidate molecules, largely selected from ICI's Company Compound Collection, using substructure search facilities of the MACCS computer programme. A simple corrosion test is used to produce a short-list of potential ligands. Each of these selected molecules is then chemically modified by introducing another functional group which is not metal-reactive, but is capable of co-reacting with the specified adhesive or polymer.

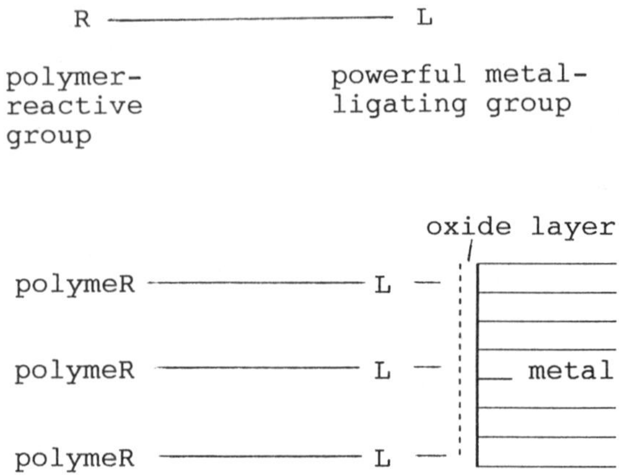


Figure 1

The effects of these surface pretreatment chemicals may be examined in several ways. Improvements in initial adhesion are easily evaluated, but positive effects on the durability of adhesive bonds often take some time to emerge even when accelerated testing is carried out.

This paper reviews some of the results generated in these recent projects. Specific details regarding structures, "fine-tuning" of adhesion promoter design etc. cannot be divulged owing to on-going technical and commercial evaluations. Questions regarding the feasibility of designing specific metal-polymer adhesion promoters, however, would be welcomed.

## 2 REVIEW OF EXPERIMENTAL DATA

Commercial adhesives and standard natural rubber formulation were used in the evaluations; cure cycles were in accordance with the manufacturers' instructions. Metal surfaces were cleaned/degreased prior to assembling the test-pieces. Simple dip, paint or meter-bar techniques were used to apply solutions of the adhesion promoters to the metal surfaces. Tests were carried out on the adhesive joints as described in the appropriate standard procedures.[3,4,5,6] Results are tabulated in Appendices 1 to 3. Loci of failure, as judged by visual inspection (usually x60 magnification) were also recorded and are given in parentheses in the tables with the following abbreviations:

C = cohesive (failure within the adhesive layer),
A = adhesive (failure at the metal-polymer interface),
(C + A)/(A + C) = mixtures of the two types, predominant type first.

Durability of epoxy to mild-steel was evaluated by storing the assembled lap-joints in distilled water at 50 C. Durability of rubber to brass was tested by exposing the rubber-encased wires to an atmosphere of saturated steam at 120 C for 24 hours.

## 3 DISCUSSION

### Epoxy Bonding to Mild-steel

In the absence of pretreatments, wet debonding is a serious problem when bonding mild-steel and a heat cured epoxy adhesive. Rapid losses in shear-strength were observed, especially in the first 500 hours of exposure. Initially the locus of failure was largely cohesive (C) but after immersion in water, failure occurred at the metal-adhesive interface (A).

Pretreating the metal with various materials which are claimed to reduce wet-debonding produced some improvements in durability (Appendix 1, table 1),and the degree of cohesive failure was increased in comparison to the controls, showing that these pretreatments were providing a degree of protection against wet-debonding. When ICI Ligand Fe/L/1 was used as the surface pretreatment, the loss in adhesive strength was at a minimum, and the locus of failure of the fractured joints was entirely cohesive within the adhesive layer (C), even after exposure to water for 10 000 hours (60 weeks). This result is taken as an indication of the complete protection of the iron oxide-polymer interface; the weakest part of the joint has been transferred to within the adhesive itself. A similar performance was observed using the ACCOMET™ C pretreatment, which is a formulation containing the highly toxic chromium(VI) species.

### Rubber Bonding to Brassed-steel

Again wet-bonding is a major problem. No improvements in initial adhesion were observed; the locus of failure was invariably within the encasing rubber, and hence the recorded adhesion is a direct measure of the tear-strength of the rubber itself. After rigorous pre-cleaning of the tyre cords and prolonged application of ligand prior to bonding with rubber, considerable improvements in durability were noted (Appendix 2, table 2). The locus of failure changed from cohesive (C) initially, to mixed (C + A), whereas control cords failed totally at the metal-rubber interface (A) after ageing.

It should be noted, however, that simple paint-applications of ligand to the tyre cords did not produce the positive effects seen when more rigorous application conditions were used. This was attributed to the presence of lubricating soaps which are resident on the cord surface. Unless these insoluble materials are removed during cleaning, they preclude the adsorption of the ligand and hence prevent it working in the intended fashion.

Acrylic Bonding to Copper

Results (Appendix 3, table 3) demonstrate the necessity of optimising the design of an adhesion promoter in order to maximise the desired effect; as the spacing distance between ligating group and polymer-reactive moeity was increased, corresponding peel strengths also improved. More synthetic work is required, but the minimum spacing distance in this system appears to be around 2 nm. It is postulated that a degree of flexibility introduced into the molecule by the 'spacer groups', can better accommodate shear stresses which occur in the adhesive during cure, in addition to allowing a "gradation of modulus" across the metal-adhesive interface. Inclusion of a "spacer group" in the design of the adhesive promoter is not always necessary; it seems to depend on the individual adhesive and metal. More research is needed before any conclusions can safely be drawn.

Selection of the polymer-reactive group also needs careful consideration; it is essential that the reactive "tail" of the molecule is compatible both physically and chemically with the adhesive, otherwise weak layers may be induced near the interface, leading to catastrophic rupture of the adhesive bond.

## 4 CONCLUSIONS

Pretreatment of a mild-steel surface with a suitably functionalised ligand, followed by bonding with epoxy adhesive ESP106 (PERMABOND™) provided a system in which the iron oxide/adhesive interface was completely protected against attack by water, inducing failure of the joint within the adhesive layer.

Significant improvements in the durability of the bonds between brassed-steel and rubber were observed, but only when the metal surface was rigorously cleaned prior to applying the adhesion promoter.

Marked improvements in the initial adhesion between copper and acrylic adhesive F245 (PERMABOND™) were measured after pretreating the metal surface with a bonding agent. Further modifications to the molecular design of this adhesion maximised the positive effect.

These three examples clearly demonstrate the potential and technical feasibility of ICI's "designer adhesion promoter" concept. It is emphasised, however, that these materials are still in the research phase of development and are not available commercially. Full development of a specific product awaits the definition of high volume/value commercial opportunities.

## REFERENCES


1. A.J. Kinloch, 'Durability of Structural Adhesives", Applied Science, London, 1983.
2. W.A. Lees, 'Adhesives in Engineering Design', The Design Council, London, 1984.
3. 'Standard Method of Test for Strength Properties of Adhesives in Shear by Tension Loading', ASTM D1002-64.
4. 'Standard Method of Test for Effect of Moisture and Temperature on Adhesive Bonds', ASTM D1151-84.
5. 'Ministry of Aviation Test Method DTD 5577', Heat Stable Structural Adhesives; Standard 90° Peel Test.
6. 'Standard Method of Test for Strength Properties of Rubber', Rubber-Wire Adhesion Test, ASTM D1871 T Method A.
7. G. Haemers, *Adhesion* (Barking, England), 1980, **4**, 175.
8. W.J. Van Ooij, *Rubber Chemistry and Technology*, 1983, **57**, 421.
9. P.A. Tasker and D. Stewart, *Construction and Building Materials*, 1988, **2**(4), 179.

## APPENDIX 1

## EPOXY BONDING TO MILD STEEL : DURABILITY RESULTS

Mild steel Q-panels (102x25x1.6mm; type RS14) were degreased in GENKLENE LV (liquid and vapour) and an aqueous solution of the ligand applied using a metering bar to give an approximate dry coverage of 50mg/m2. Single-lap joints were assembled using an epoxy single-pack adhesive (ESP 106 ex PERMABOND Ltd). the joints were cured under light pressure at 180 C for 1 hour.

TABLE 1

| PRETREATMENT | SHEAR STRENGTH MPa / LOCUS OF FAILURE<br>IMMERSION TIME IN WATER AT 50 C | | | | |
|---|---|---|---|---|---|
| | 0hr | 500hr | 1500hr | 3000hr | 5000hr |
| None | 39.4<br>c+a | 20.3<br>a+c | 20.1<br>a+c | 17.0<br>a | 13.8<br>a |
| 1% aq soln of glycid-oxypropyltrimethoxy silane (dip applicn) | 39.4<br>c+a | 33.1<br>c+a | 29.1<br>c+a | 23.4<br>c+a | --- |
| Citrate clean, dip in methanol soln of dibenzoylmethane | 38.8<br>c+a | 28.5<br>c+a | 22.1<br>c+a | 21.3<br>c+a | --- |
| Citrate clean, dip in methanol soln of 1,2,6-hexanetriol-trithioglycolate | 38.8<br>c | 29.7<br>c+a | 28.7<br>c+a | 25.1<br>c+a | --- |
| Accomet C (ex Albright & Wilson) | 42.8<br>c | 36.2<br>c | 30.7<br>c | 28.6<br>c | --- |
| ICI LIGAND Fe/L/1 ca 50mg/m2 | 43.3<br>c | 31.9<br>c | 30.2<br>c | 29.2<br>c | 24.3<br>c |

## APPENDIX 2

## RUBBER BONDING TO BRASSED STEEL : DURABILITY RESULTS

Single strand brassed steel tyre cords (0.35mm)were degreased in PERKLONE (liquid and vapour, repeated 5x) and immersed in an aqueous 1% solution of the ligand for 1 hour at 50 C then dried. The wires were embedded in a natural rubber formulation (ex Vulnax Int) and cured at 140 C for 10 mins.

TABLE 2

| PRETREATMENT | ADHESION (N) / LOCUS OF FAILURE | |
|---|---|---|
| | INITIAL | AFTER AGEING * |
| None | 230<br>c | 85<br>a |
| ICI LIGAND<br>Zn/Cu/L/1 | 235<br>c | 165<br>c+a |
| ICI LIGAND<br>Zn/Cu/L/2 | 245<br>c | 160<br>c+a |

* 24 hr at 120 C

## APPENDIX 3

## ACRYLIC BONDING TO COPPER FOIL : INITIAL ADHESION

Rigid copper-clad laminate (0.5mm ex Prestwick Circuits plc) was cleaned (wet abrasive scrub) and degreased in GENKLENE LV. The ligands were applied by painting a dilute alcoholic solution onto the metal surface, to give an approximate dry surface coverage of 50 mg/m2. Two such strips were adhered using an acrylic adhesive (Permabond F245)

TABLE 3

| PRETREATMENT | SEPARATION L ____ R | 90 deg PEEL N/mm | PEEL MODE deviation |
|---|---|---|---|
| None | ---- | 1.7 | + 20% |
| Cu/L/1 | none L-gp only | 1.5 | +20% |
| Cu/L/2 | none R-gp only | 1.7 | +20% |
| Cu/L/3 | 7A | 2.2 | +20% |
| Cu/L/4 | 11A | 3.0 | +15% |
| Cu/L/5 | 14A | 4.0 | +15% |
| Cu/L/6 | 20A | 4.9 | +7% |
| Cu/L/7 | 23A | 4.5 | +10% |

# ENVIRONMENTALLY ACCEPTABLE COATINGS FOR THE AUTOMOTIVE INDUSTRY

J.L. Pearson and A. Frangou

IDAC UK Ltd
Slough
Berkshire, SL2 5DS
United Kingdom

## 1 INTRODUCTION

Over the last decade, the Automotive Industry has been the focal point for much technological innovation and change. Innovation in component design, in processing and materials, and in the design of the vehicle itself. Perhaps the major impetus for these advances is the continuous quest for improved quality and performance, demanded because of:

a) increased customer awareness and expectation,
b) competitive pressure from both inside and outside the European domestic market, particularly Japan,
c) increased utilisation of vehicle design and appearance as a major advertising and selling feature.

During this period, both finishing processes and the coatings used have become highly sophisticated, some of the more significant trends being:

a) increased usage of zinc coated steels,
b) adoption of dip phosphating processes,
c) almost universal conversion to cathodic electropaint for all vehicles,
d) all metallic, as well as increasing numbers of solid colours, being processed using the basecoat-clearcoat process,
e) greater degree of automation in plants, with all spray coatings being applied

electrostatically.

All the above, although adding to the complexity, are designed to provide improved quality and performance; in particular to provide uniform quality.

However, during this period, other forces have increasingly affected the industry. The first of these is the responsibility of employers to ensure the safety and welfare of all its employees via the Health and Safety at Work Act (HASAWA). Hence, for example, in the United Kingdom, the usage and spraying of isocyanate containing paints is controlled and monitored. Such controls dictate and limit the chemistry the paint formulator has at his disposal in his search to provide the highest quality products. The second of these is the increasing global concern for the safety of the environment. Today, legislation is in place in most parts of the industrialised world which controls the release of toxic substances to both water and air.

The Automotive Paint Industry has two major challenges. Firstly, pretreatment chemicals and anti-corrosive electrocoat primers are formulated using heavy metals such as lead and chromium, which poison water supplies. To eliminate these the industry has the following product and process goals.

a) increased usage of zinc coated steels,
b) full dip pretreatment process,
c) non-chrome containing passivation rinses,
d) lead-free electropaints using alternative *non-toxic* corrosion inhibiting pigments.

Secondly, spray applied automotive finishes are traditionally formulated at low solids contents and contain large quantities of hydrocarbon which when emitted in either virgin or oxidised forms cause environmental damage. These chemicals must be reduced and, eventually, eliminated if the industry is to play its part in the protection of the world's environment.

Although both of the above are of major importance and concern, this paper will concentrate on atmospheric pollution.

## 2 ANTI-POLLUTION LEGISLATION

Already, as a result of pressures from environmentalists, legislation exists which limits the emission of volatiles to the atmosphere, both in type and quantity. Major legislative thrusts are in the USA, Germany and Scandinavia; however, all western car producing countries are becoming increasingly aware of the problem, such that the timing of more unilateral legislation becomes the only uncertainty.

Legislation in the USA

In the USA, solvent contents are measured in terms of volatile organic content (VOC) expressed as pound of solvent per US gallon of wet paint as applied. The Environmental Protection Agency (EPA) has set a target of 2.8 lb for all coatings, a major reduction from the 6 to 7 lb of many existing systems.

In an attempt to meet this target, the North American paint industry has developed high solids coatings in both the primer and topcoat areas using in many cases flexibilised, low molecular weight versions of existing resins, and cross-linking these with low viscosity monomeric melamines. A further 'trick' is to apply the coatings at elevated temperature (40 C), thus reducing application viscosity and allowing solids to be further increased. This trend has led to a reduction in VOCs, such that values for commonly used finishes are shown in table 1.

Table 1 Analytical VOCs of Typical Automotive Finishes

| Product | VOC (lb/US gallon) |
|---|---|
| high solids primer surfacer | 3.5 |
| conventional metallic basecoats* | 5.2 - 6.8 |
| water-borne metallic basecoats | 2.8 - 3.0 |
| high solids metallic basecoats | 3.8 - 4.0 |
| conventional clearcoats | 4.2 - 5.5 |
| high solids clearcoat | 3.5 - 3.7 |

(* basecoats in the solids range 12 to 20% by weight)

However, because the total solvent associated with any particular painting operation is a function not only of the paint used, but also of the painting operation, this concept has been extended to describe the VOC of one gallon of applied solids, where

$$\text{VOC/gallon applied solids} = \frac{\text{VOC}}{\text{(volume solids)} \times \text{(transfer efficiency)}}$$

This definition, therefore, is a function of both paint and process. In order to complete this calculation, the EPA has assigned transfer efficiency values, based on experience and measurement, for the various painting operations as shown by table 2.

Table 2 Assigned Transfer Efficiencies for various Painting Operations

| Painting Operation | Transfer Efficiency (%) |
|---|---|
| air atomised water-borne | 30 |
| air atomised solvent-borne | 40 |
| manual electrostatic | 62 |
| automatic electrostatic | 75 |
| electrodeposition | 100 |

Using these values, several basic definitions and targets can be defined, these being the basis for the emission control of automotive plants in the USA.

Reasonably Achievable Control Technology (RACT). Assigned a value of 15.0 lb, is the target for existing facilities. This is achievable using state of the art product technology.

New Source Performance Standards (NSPS). Assigned a value of 12.2 lb, is the allowable maximum for new facilities. This is achieved by combination of state of the art products and application techniques.

Lowest Available Emission Rate (LAER). This has recently been introduced to allow authorities the power to demand lower values than NSPS. This value will incorporate not only product and application technology, but also solvent incineration, absorption etc. This is the practical goal any new facility may be requested to meet.

As targets are gradually reduced, the high solids systems cannot follow these demands, and there is a greater awareness of the possible advantages of water-based products, which not only meet existing targets but, as the technology is advanced, present the opportunity for further reduction.

## Legislation in Europe

The most advanced anti-pollution regulations in Europe exist in Germany, known as "T.A. Luft". This is a Government Directive and applies to authorising bodies so that they apply uniform standards and treatment to all plant operators.

In order to comply with this legislation it is important to understand the "Mass Flow" of emissions from the entire plant, as specific regulations apply to:

a) total emissions from the entire plant,
b) the concentration of organic volatiles in the exhaust air from spray booths,
c) the concentration of carbon exhausted from oven stacks.

Of particular significance and importance is the limit for the total solvent that can be associated with the entire manufacturing process.

For vehicles finished in a metallic colour, the maximum permitted solvent level is 120 g $m^{-2}$ of total surface (internal and external area). For example, a typical medium-size vehicle has a total surface area of around 70 to 80 $m^2$, hence the total solvent which can be associated with the production of such a vehicle would be 8.4 to 9.6 kg. For vehicles painted in a solid colours this maximum is 60 g $m^{-2}$.

However, T.A. Luft states that "The possibilities for further reducing solvent emissions through either the use of low solvent (or solvent free) paint systems *or* by waste gas cleaning, especially in spraying areas, must be fully examined." In other words, the future targets of solvent levels will, as in the USA, depend on state of the art technologies, combinations of products and processing not either/or.

Legislation in the United Kingdom

Recently a draft document has been issued by H.M.I.P., which is very similar in format and content to T.A. Luft, in as much as it provides guidelines for authorising bodies, as well as defining maximum allowable levels for solvent emissions. Once again, these limits are based on the best available technology in terms of both coatings and processes, making specific mention of the following:

a) the use of water-based paints,
b) the use of electrostatic spraying,
c) the reduction of exhaust volume,
d) recirculation of exhaust gases coupled with more automation of paint application,
e) the use of higher solids content paints.

The strict control of solvent emissions for the United Kingdom automotive plants is "just around the corner".

## 3 THE EUROPEAN AUTOMOTIVE FINISHING PROCESS

Correctly applied, the products currently used by European automotive manufacturers provide the highest standard of appearance available on mass produced vehicles. Whereas, in North America, there have been moves to reduce the number of coats, for example, elimination of primer-surfacer, this trend is not being followed by the European industry (or the Japanese). The well established multi-stage process is shown in table 3.

Table 3 European Finishing Process

| Coating | Composition and method of application |
|---|---|
| pretreatment | spray or, preferably, dip applied inorganic layer to provide corrosion protection and a key for the electroprimer. |
| cathodic electrocoat | epoxy based electrodeposited primer which imparts uniform corrosion protection over the whole body. |
| primer surfacer | spray applied polyester (or urethane modified polyester) imparting:<br>- filling properties,<br>- smoothness,<br>- sandability, for ease of removal of small defects,<br>- mechanical strength (chip resistance). |
| topcoat | durability and cosmetic appearance. |

Today, virtually all passenger vehicles painted in metallic and mica colours (as well as increasing numbers of solid colours) are finished using the basecoat-clearcoat process, a thin polyester-CAB based pigmented colour coat, followed by wet on wet application of a thicker (35 to 40 micron) acrylic-MF clearcoat. This system provides a combination of maximum durability and performance with styling capability, and will form the basis of the remainder of this article.

### Total Emissions from the Basecoat-Clearcoat Process

Let us now consider the process previously described, including conservation wax which is normally applied to the vehicle to provide transport and storage protection. These waxes are carried in solvent and provide a further source of emissions. A typical process therefore becomes:

a) electroprimer (22 to 25 microns),
b) primer surfacer and stone chip surfacer (30 to 35 microns),
c) basecoat (10 to 15 microns), higher builds for some solid colours and poor hiding metallics,
d) clearcoat (35 to 40 microns),
e) conservation wax.

All sprayable coatings are applied electrostatically, with the exception of the metallic basecoat which is applied in two coats, 50/50 electrostatic/air atomised. The total solvent associated with such a process lies in the region of 130 to 150 g $m^{-2}$, relatively apportioned as shown in table 4.

Table 4 Relative Proportion of Solvents Associated with each Process Stage.

| Process stage | Proportion of solvent associated with the process stage (%) |
|---|---|
| electropaint | 1.6 |
| primer surfacer & stone chip primer | 16.1 |
| metallic basecoat | 51.6 |
| clearcoat | 19.4 |
| protective wax | 11.3 |

These are average figures extrapolated from various sources, which provide the following information.

a) The electropaint, a water based coating accounts for a very small percentage of the total. Nevertheless, new generation coatings are now being introduced which reduce solvent and volatile levels by as much as 50%.

b) Approximately 50% of the solvent associated with this process emanates from the metallic basecoat. It is not surprising, therefore, that a high proportion of research and development effort has been concentrated on the production of low emission basecoats.

c) The remainder is split between the primer surfacer, clearcoat and wax, each making a substantial contribution and a potential major saving.

## 4 METALLIC BASECOATS

One of the major features of a vehicle finished in a metallic base-clear system is the change in hue and brightness when the coating is viewed at different angles. This effect is maximised when the film contains a high ratio of aluminium flakes to prime pigment (which must be transparent) and when all the aluminium flakes are aligned parallel to the substrate. Under these circumstances the film acts as a mirror.

Let us consider light shone onto the film at an angle near normal to the surface. When viewed normal to the substrate the film appears bright as most of the light is reflected backwards. This is described as displaying high "head on brightness". If the film is now turned and viewed at a very low angle it appears dark, as no light is reflected back along this angle. This is known as exhibiting high "flip-tone" or "flop effect".

The highest standard of metallic effect in conventional solvent-borne technology is achieved using polyester basecoats applied at gun solids around 10 to 12% by weight. The level of CAB and waxes associated with this general product type provide maximum formulation and application latitude. In particular, they control the viscosity profile during and immediately following application, preventing aluminium flake movement in the wet film. This, in combination with the very high degree of film shrinkage during drying, produces uniform, parallel alignment.

As the solids content of the basecoat is increased two factors come into play. Firstly, the degree of film shrinkage is greatly reduced, and secondly, viscosity control is much more difficult to achieve. Formulating at higher solids content means resin molecular weights are lowered, and the level of CAB reduced (or eliminated). This presents two possible problems. Firstly, immediately following application the evaporation of solvents from the film causes local "eddy currents" which produce aluminium movement, and secondly, after clearcoat application penetration or, strike-in of the clearcoat into the basecoat can occur. All these factors combine to reduce both head-on brightness and the flop effect. Figure 1 shows the effect of weight solids on the brightness index of silver basecoats, all formulated using the same aluminium grade.

In practice, it is found that the maximum usable weight solids before metallic appearance is degraded is 18 to 20%, this being substantiated by the data presented in figure 1. At solids far in excess of this value, for example, the 45% of North American metallic basecoats, the result is a grey and grainy appearance. In order to compensate for this effect, new aluminium grades have been developed, however, the major trend has been for the North American industry to find alternative forms of glamour effect, either via the use of coarse, high sparkle flakes or the move to darker, richer colours formulated using mica pigments which do not rely on low solids and film shrinkage to optimise appearance.

The degree of aluminium flake orientation can be measured using a goniophotometer, as shown in figure 2. An incident beam is shone onto the clearcoat film covered by a glass hemisphere with an intermediate oil film to eliminate specular reflection, at an angle of 45°. The reflected light is then measured using a photocell at various angles of reflection. If all flakes are parallel then all light will be reflected at 45° angle of reflectance. The resultant graph of light intensity against reflectance angle gives, therefore, a measure of the uniformity of flake alignment. Figure 3 shows two curves, curve "A" shows a sharp peak around 45°, and would be typical for a low solids system. Curve "B" is much less defined with more uniform scatter, and would be typical of a high solids system. Cross-sectional micrographs show the flakes in "B" to be much more randomly oriented, the film appearing greyer than "A" with much less degree of flip. Although the high solids technologies impart certain styling limitations, they do reduce emissions by as much as 70 to 80%.

## 5 WATER-BORNE BASECOATS

Theoretically, it is possible to retain low solids, along with its previously discussed advantages, and still reduce solvent emissions if the solvent is replaced by water. The use of water-borne automotive topcoats is not a new concept, General Motors having pioneered such a system in the late 1970s in the USA. However, their system was a one coat topcoat, not a basecoat, and the "mark 1" technology on which it was based required strict humidity and temperature control, as well as multiple coat application.

The concept of a basecoat was pioneered in Europe, initially by ICI (one of IDAC's parents), and then adopted by DuPont in the USA (the other IDAC parent). In understanding this concept it is important to understand how a solvent-borne basecoat is applied. During application a liquid coating requires low viscosity at high shear rates to achieve satisfactory atomisation in the spraygun, and pumpability in the circulation system. Immediately after application, the coating on the substrate must have a high viscosity to avoid sagging or aluminium/pigment flotation. In a solvent system this is achieved by the rapid evaporation of low boiling solvent. The mark 1 water-based technology previously mentioned attempted to orchestrate this viscosity change using small amounts of low boiling co-solvents. Unfortunately, if the coating is to retain a

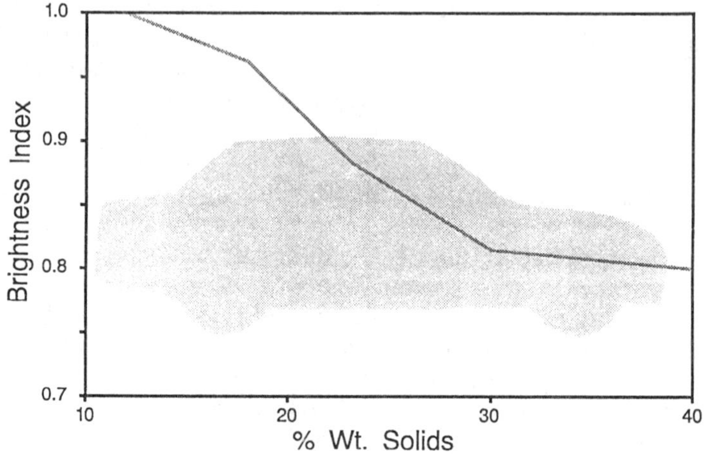


Figure 1 Brightness Index as a function of Solids

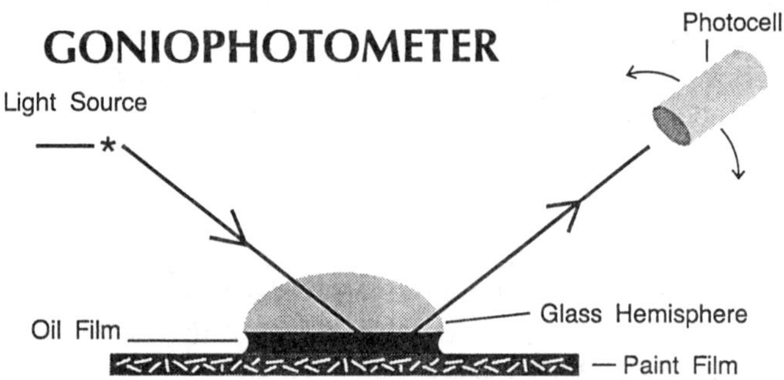


Figure 2 Measurement of metallic flake orientation

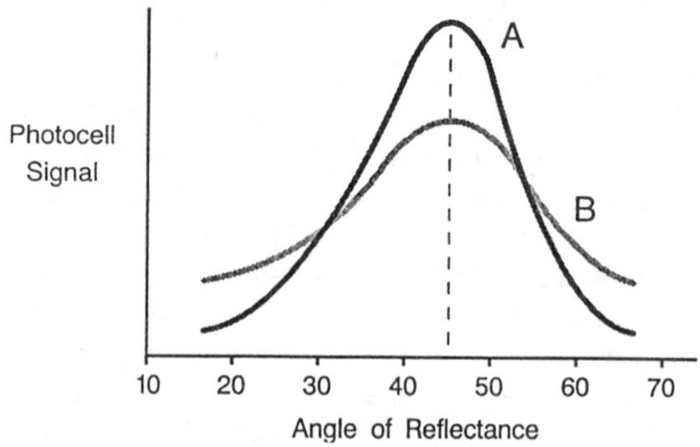


Figure 3 Goniophotometric curves

definition of being water-based, the level of co-solvent available to provide this function is insufficient to give a satisfactory application window.

It has been found by several workers that compensation for loss of control of viscosity by solvent evaporation can be achieved by producing systems which exhibit pseudoplastic rheological characteristics, as shown by figure 4. At high shear rates, viscosity is low, allowing ease of atomisation and circulation, however, as the shear rate decreases the viscosity increases immediately, without any time depenence, that is , non-thixotropic, and does not rely on co-solvent or water evaporation to achieve this increase. In practice, during the spraying process some water and co-solvent is lost,thus giving rise to a further increase in viscosity; these advantages of pseudoplastic behaviour are:

a) excellent anti-setting properties in the container or dead spaces in the circulation system,
b) excellent flake control,
c) excellent sag control,
d) greater temperature/relative humidity application window,
e) excellent atomisation properties.

Paints formulated to give this viscosity profile can be characterised by measuring viscosity as a function of shear rate using, for example, a Contrave's rheomat. A true pseudoplastic system will obey Ostwald's power law,

$$\tau = \kappa\gamma^{n}$$

where $\tau$ = shear stress, $\gamma$ = shear rate, $\kappa$ = constant, n = viscosity index.

A log/log plot of shear stress against shear rate should produce a curve as shown by figure 5. The value of $\kappa$ is equal to $\tau$ when $\gamma = 1$, and is equal to the extrapolated value of the viscosity at a shear rate of 1 $sec^{-1}$, known as the low shear viscosity (LSV). In practice this is measured using a Brookfield viscometer under specified conditions. This provides an easy control measurement "in the field". A further characterisation point is the high shear viscosity (HSV) measured using an ICI cone and plate viscometer, once again allowing plant checks to be made easily. Typical values for metallic basecoats would be:

LSV = 20 Poise @ 1 $sec^{-1}$
HSV = 0.30 Poise @ $10^4$ $sec^{-1}$

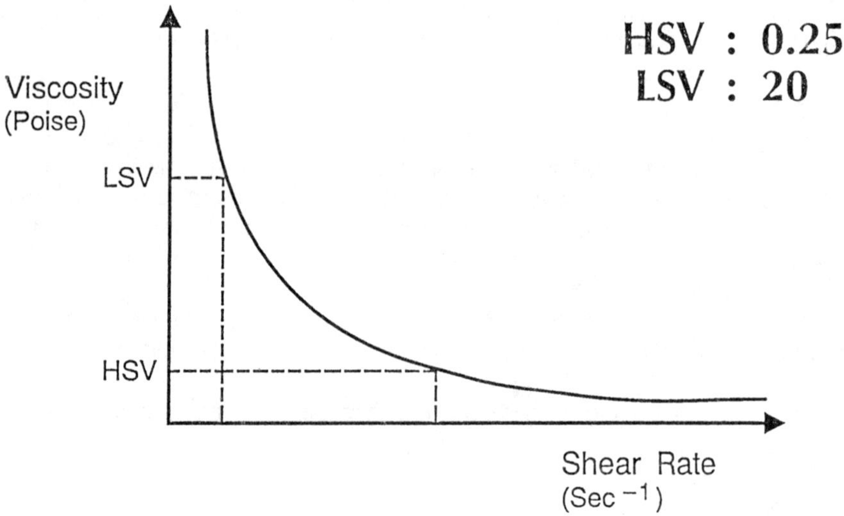


Figure 4 Pseudoplastic behaviour

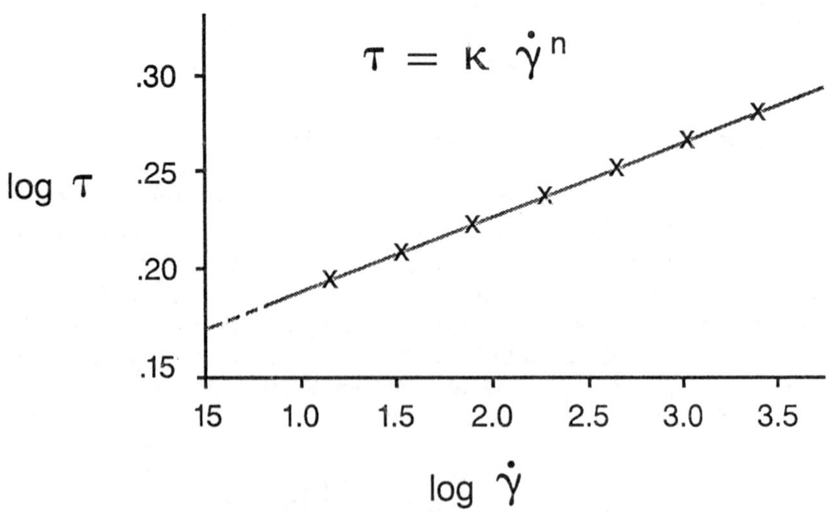


Figure 5 Basecoat characterization

## Aqueous Microgels

The key to the success of Aquabase™ water-borne basecoats is the aqueous acrylic microgel polymer dispersion which imparts the required pseudoplastic characteristics, as well as providing mechanical strength and resistance to attack by clearcoat. The microgels can be characterised as shown in figure 6. Each particle consists of a cross-linked acrylic core, this cross-linking imparting mechanical strength and is insoluble in both water and solvents. This core is encapsulated with a mantle of acrylic polyelectrolyte containing both hydroxyl and carbonyl functionality. In its unneutralised form this latex exhibits Newtonian, low viscosity behaviour as the mantle is collapsed. On neutralisation with base the pseudoplastic rheological characteristics are realised, and when formulated into a basecoat the gel particles can be cross-linked via the hydroxy groups with melamine formaldehyde and other water soluble cross-linkers. In summary, this core/mantle morphology provides the following properties/characteristics:

a) the core provides
- resistance to solvent attack,
- mechanical strength,
- film durability,
- chemical resistance.

b) the mantle provides
- adhesion properties to substrate and clearcoat,
- cross-linking potential,
- stability, both static and dynamic,
- solubility and rheological properties.

Basecoats are normally formulated at solids levels of 16 to 18%, with 12 to 14% co-solvent. As such they combine the styling aspects associated with low solids solvent-borne systems with the low emission levels of the highest, high solids systems.

## Production Line Experience

Water-borne basecoats have been commercial in both North America and Europe since 1986, with almost one million vehicles finished in Aquabase, confirming the validity of the concept previously described. It has been found that water-borne basecoats apply in an analogous way to solvent-borne basecoats with the following modifications or constraints.

a) Extra energy is required for adequate atomisation owing to its rheological

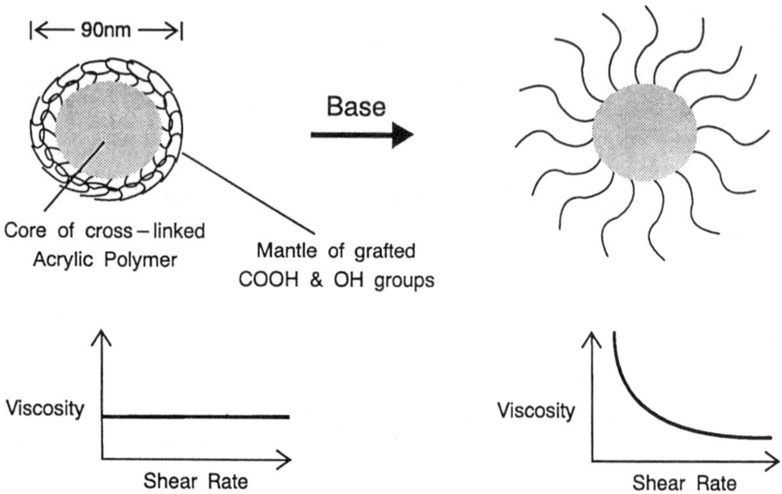


**AQUEOUS ACRYLIC MICROGELS**

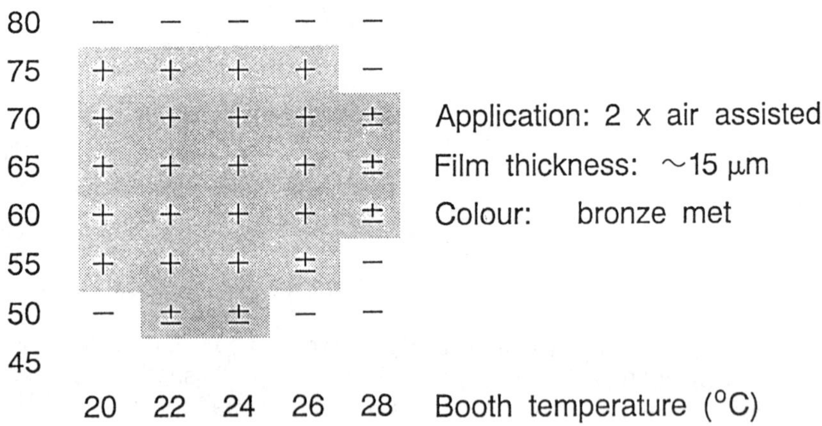


Figure 7 **INFLUENCE OF CHANGES IN TEMPERATURE AND RELATIVE HUMIDITY ON COLOUR**

characteristics. This can be achieved by:

- minimum air pressure of 5 to 6 Bar at the gun,
- limiting fluid deliveries to 400 to 600 cm $min^{-1}$ depending on colour,
- for high line speeds use twin gun set-ups.

b) A drying zone of three minutes comprising one minute infra-red and two minutes warm air is required to dry the basecoat prior to clearcoating.

c) All parts which come into contact with water-borne products must be made from austenitic stainless steel, including:

- storage containers,
- mixers,
- filters,
- circulation pumps and pipes,
- spray guns.

d) Although Aquabase has a much wider application tolerance than conventional water-borne technology, there are still temperature and RH constraints, as shown by figure 7.

## 6 LOW EMISSION PRIMERS AND CLEARCOATS

In the area of primer surfaces and clearcoats high solids solvent-borne systems have gained much universal acceptance; however, technologies have almost reached a plateau. For this reason water-based and powder systems are now receiving more attention.

### Primer-surfaces

Water-borne first generation products are presently used successfully by one major car company in Europe and are being slowly adopted by others. Major problems are associated with adhesion, particularly certain electrocoats. Second generation products are being generated based on more sophisticated polyester and polyurethane resins. At solvent levels of 5% or less this type of product is highly attractive from an environmental standpoint, and are the most likely future standards for the industry.

As stoving temperatures for primer surfaces are high (>160 C), the use of powder surfaces remain a possibility. However, whether other constraints associated with this generic product type can be overcome and provide advantages over water-borne systems is debatable. As always with paint systems it is probable that both water and powder will find a place in the market.

Clearcoats

Here, the major requirement is for improved properties, including scratch resistance, resistance to acid rain and increased durability. For this reason the industry is moving to a new generation of clearcoats which include 2K urethane, 2K non-isocyanate, fluoroclears.

# AUTOMOTIVE ANTI-CORROSION PRETREATMENT;

## Trends and Controls of Pretreatment Systems for the Automotive Industry

D. Galjaard

Novamax Technologies
Brussels
Belgium

In the worldwide automotive industry, all manufacturers are striving continuously to improve the life span of their products. The improvement in cosmetic and perforation corrosion is a major target helped enormously by the introduction of cathodic electropaint about twelve years ago. This introduction has led to greatly improved and novel metal pretreatment processes which have increased this benefit and achieved a major goal in higher corrosion protection. The prevention of perforation has also been improved by the use of precoated steels such as electro-zinc, single or double sided, hot dip galvanised, other alloyed electro-zinc such as zinc-nickel and zinc-iron and duplex powdered zinc systems. How the automotive industry, working together with its suppliers, operates and controls its pretreatment systems to achieve maximum quality and performance will now be outlined and a look taken at future trends.

In the early 1980s the view in the industry became firmer that there was a significant benefit in changing from spray zinc phosphate processing to dip zinc phosphate processing which had two major benefits; a) improved internal coverage, b) vastly improved "cubic" crystal structure. This was coupled with the availability of low zinc processes which incorporated zinc iron phosphate (phosphophillite) into the traditonal zinc phosphate (hopeite) coating. The introduction of coated steels into the car body structure developed the need and introduction of multi metal zinc phosphate processes which used nickel and manganese in their formulation and the subsequent coating. This trication zinc phosphate system has been found to give greatly improved salt spray and scab corrosion protection to the painted vehicle.

These processes produce a cubic-type crystal structure on cold rolled steel and include phosphates of zinc, nickel, manganese and iron in their composition. Because of their high alkali solubility resistance they have greatly increased the corrosion resistance of the painted car body. On the zinc coated and aluminium surfaces a hopeite structure of very fine needle crystals parallel to the surface is produced which are highly resistant to corrosion and have excellent adhesion properties for the applied cathodic electropaint.

Automotive manufacturers with spray phosphate machines have looked at reproducing dip phosphate conditions by flooding the internal section of the vehicle and the strategic use of coated steels to provide the quality of performance necessary in today's market. This, in addition, benefited from the much higher corrosion protection offered by the zinc phosphate processes.

The higher quality and corrosion protection standards now present in the industry have been brought about by the chemical supply companies responding to the market need. However, it has brought additional demands in the control of the materials and the process plants that treat the car bodies. The mechanical properties of the total paint system are reliant on the adhesion properties of the phosphate coating and the total paint system. These are tested by use of an impact test, cross hatch tape test and bend tests. The important characteristics of the phosphate coating for these properties are coating weight and crystal size and orientation, all controlled on line. Coating weights are usually in the range of 2 to 4 g $m^{-2}$ and the preferred crystal structure is cubic and 1 to 5 microns in diameter determined by SEM. The orientation of the crystals can affect the "wet adhesion" test results of the total paint system; these are also controlled on line with a regular off-line SEM evaluation for crystal size, orientation and composition. The corrosion characteristics of the phosphate coating are measured by cyclic corrosion, filiform, exterior and test track corrosion tests on a regular basis. The phosphating factors which have an effect on this important property are the alkaline solubility resistance of the coating, the nickel and manganese content of the bath and the P-ratio of the coating. These are all controlled by regular testing.

Regular and frequent control testing of the process solution is made to ensure consistency of operation and results. Additional monitors are suppliers' reports to the user manufacturers to ensure that the highest standards of control are in place. These reports will incude:

a) weekly quality audit,
b) weekly line condition report,
c) weekly maintenance report,
d) monthly cost/usage,
e) monthly test panel audit,
f) monthly solution analysis audit,
g) monthly quality audit summary (SQC).

Great importance is placed on crystal structure and orientation and is monitored all over the body at predetermined sites at regular intervals. Any deviation from the required size, structure and orientation of the crystal is corrected by the necessary adjustments to the process parameters. A marked variation in coating structure can occur with test panels of different age and source and, although, they may be a guide to good performance it is necessary to be able to check the coating characteristics of the processed car body.

The recent improvement in corrosion resistance determined by pretreatment factors can be summarised as:

a) strong emphasis towards dip processing,
b) extended use of zinc, zinc alloy and aluminium substrates,
c) introduction of advanced transition metal phosphate coatings,
d) trend to computerised bath control.

How can we look to futher improvements in the future?

a) Immersion processing will increase as the need to replace or install new processing facilities occurs.

b) The trend towards the use of coated steels is proceeding in different directions depending on the global area and the particular manufacturer (and to some extent availability of the chosen substrate). In Japan, steel has been partly replaced by hot-dip galvanised and electro-zinc coated steel but whilst these will continue in use the tendency is towards 'zinc alloy' coated steels and some duplex coatings. This trend, in general, is being followed by 'transplant' operations in the USA and Europe. In Europe the main use is with hot dip galvanised, electro-zinc and zinc-nickel alloy with current trend possibly towards 'refined' (smoothed) hot-dip galvanised and some duplex materials. North America is fairly stable using hot-dip galvanised and electro-zinc.

c) Introduction of transition metal phosphate coatings has considerably uprated corrosion performance related to the phosphate coating. It is likely that this type of technology will provide the main line phosphate technology for the early 1990s. There is still room for process development to refine its capabilities and improve control and performance. New or modified cleaning formulations will emerge and the possibility of an improved surface conditioner and non-chrome passivation compound must be seriously expected. The effect of environmental legislation is another incentive which may force supplier companies towards newer, inventive products. Various routes to short and long-term improvement goals are set out in tables 2, 3 & 4.

d) The trend to more computerised pretreatment control continues to advance. As techniques become available they will be adopted and introduced to improve the already high standards being achieved (table1).

The automotive manufacturing companies and their pretreatment chemical suppliers face a big challenge in the 1990s. The 1980s saw a massive gain in anti-corrosion properties with the introduction of transition metal phosphate coatings substantially improving the corrosion performance of cathodic electropaint primer and the total paint system. This introduction also complemented the increased use of coated steels to prevent further perforation corrosion and eliminate the red rust corrosion aspects of paint chipping. We will undoubtedly see a closer co-operation between metal pretreatment suppliers, paint suppliers and the automotive industry in the future.

## APPENDIX 1

### AUTOMOTIVE ANTI-CORROSION MANUFACTURING AND PROCESS DEVELOPMENT

| Demand | Response | Action |
|---|---|---|
| up-rating to 10 year corrosion prevention warranty in North America and Europe | promotion of full-dip system | establishment of equipment specification |
| use of zinc-plated steels | counteract possibility of white spotting and poor wet adhesion | development of low $SiO_2$ or non-$SiO_2$ cleaners |
| modification/improvement of cathodic ED | improved resistance to coating dissolution at the high pH at electro-coating site | development of Mn-modified phosphating chemical |
| use of continuous-annealed BH steel | improve process for variable steel sheet surface conditions | development of cleaner and surface conditioner |
| energy saving | establishment of lower temperature operation | development of lower temperature cleaning/phosphating chemicals |
| consistency of line operation | improve surface conditioning | study on $Mg/P_2O_7$ ratio |
| higher quality by improving paint film appearance | evaluation of iron filing removal system | installation of suitable filtering system |

## APPENDIX 2

## RUST PREVENTIVE MEASURE IN RESPONSE TO VARIOUS MODES OF CORROSIVE ATTACK

| Corrosion type | Material | Car body engineering |
|---|---|---|
| cosmetic | precoated steels (2-sided) | car structure less subject to chip damage |
| | aluminium | smoothing of edges and lapped joints |
| | plastics | |
| | full-dip phosphating | improved trim fit technology (plastic inserts trim strip adhesive fixing) |
| | stress relaxation type CED | |
| | high ED film (edge covering) | |
| | intermediate coat for chip resistance | |
| | higher total paint film | |
| | improved seam sealing | |
| perforation | precoated steels | moisture/salt mud resistant body |
| | plastics | |
| | full-dip phosphating and cleaning | avoid air pockets in phosphating, electro-plating etc. |
| | CED | plastics liner |
| | sealer | avoid difficulties caused by edges and burrs |
| | rust-preventive sealant, PV sol | |
| | rust-preventive wax | |

| Corrosion type | Material | Car body engineering |
|---|---|---|
| | thick coating film | |
| | post galvanising low volume sub-assemblies | |
| galvanic | full dip phosphating | plastics, rubber spacers |
| | CED | adhesive fitting for model badges |
| | thicker film as total coating | improved trim fit |

# APPENDIX 3

## PRESENT TECHNICAL TREND IN THE AUTOMOTIVE INDUSTRY

<u>Extension of the quality assurance term</u>

| | | | |
|---|---|---|---|
| high corrosion resistance (perforation) | 5 years ——— | 7 years ——— | 10 years |

<u>Expected changes in materials satisfying above requirements</u>

| | | | |
|---|---|---|---|
| base metal | steel ——— | high strength steel ———<br>galvanised steel | electro alloy steel |
| cation ED | cation ED ——— | ——— | thick CED coating |
| phosphate chemical | low zinc ———<br>zincolene 600EC | low zinc - full dip ———<br>zincolene 603/613 | Mn-Zn phosphate coat process<br>zincolene 616/617 |
| characteristics of coating | phosphophyllite ——— | new phosphophyllite ——— | fine crystal new phosphophyllite |

## APPENDIX 4

## FUTURE DIRECTIONS

### <u>Statistical process control</u>

computerised - on line data base
upper/lower control limits

### <u>Computerised line control</u>

programmable controllers
modems - supplier control centre
centralised control centre
networking

### <u>Computerised bath control</u>

multi-replenisher products
 zinc/pH interactive effects
 dissolved oxygen/hydrogen content

### <u>On-line coating analysis</u>

coating density, weight, analysis
coating quality controlling line process

# CORROSION PROTECTION OF VEHICLES

M. Gribby

Mebon Ltd
Huthwaite
Sutton-in-Ashfield
Nottinghamshire, NG17 2RL
United Kingdom

## 1 INTRODUCTION

A new coach or bus is an expensive asset for any operator to purchase. Such purchases are not normally made without considerable forethought as the vehicle will normally be considered as an item of considerable prestige. The value of a coach or bus will be very dependent on its longevity and in particular its eventual re-sale value. Traditional coatings used to protect steel frames of vehicles have allowed premature structural failure or have required periodic re-application, both of which are unpredictable and costly.

Underbody corrosion is of particular concern to many operators where the limitations of conventional barrier systems such as bituminous coatings or wax based coatings has led to costly maintenance or early scrapping. Advances in coatings' technology such as stress relieved films, new anti-corrosive mechanisms and tolerant application, all contribute to the achievement of more reliable and durable systems. Coupled with realistic methods of predicting performance, longer life systems can be specified with greater economic certainty providing cost effective benefits to coach and bus operators.

## 2 TRADITIONAL SYSTEMS

Bituminous and wax based coatings have been a popular method of protecting vehicle steelwork in the past, but both these products have certain disadvantages.

Bitumen based coatings are effective water barriers and are often applied as very thick coatings in the belief that thicker coatings are better, the demise of these coatings is in that they do not function as a true anti-corrosive system. Bitumens will collect contamination when warm and become extremely brittle when cold or aged, which results in cracking of the film and also loss of adhesion. As bitumen materials do not prevent the spread of corrosion under the film, corrosion will often occur and be disguised in its early development by the sheer dry film thickness of the coating. When corrosion is obvious, it is unfortunately then too late to repair it. This is further complicated by there being very few coatings which are compatible with bituminous materials.

Wax based materials have good adhesive properties to most of the substances which are used in the manufacture of coaches or buses because the material is a soft flexible coating. Unfortunately, wax based coatings do not have resistance to mechanical damage which will be caused by stone chips and wheel water wash. This results in the erosion of the coating after a relatively short time. Another factor which compounds this problem is the increasing use of steam cleaning prior to vehicle inspections or repair, heat from the steam cleaner will soften the wax and the pressure from the water wash will often clean away the wax from the steel substrate. These factors mean that if wax based products are used to protect a coach or bus, operators should ensure that the vehicle is inspected annually and re-treated with the wax coating to ensure that no corrosion occurs.

The increasing use of road salts, figure 1, in cooler climates and the higher expectations of the vehicle operator with regard to daily operation and improved performance of the vehicle, have all resulted in the need for a system to incorporate an effective electrochemical inhibitor which interferes with the natural corrosion process.

## 3 PERFORMANCE REQUIREMENTS OF HIGH PERFORMANCE PROTECTIVE COATINGS

There are three key factors which determine the durability of any vehicle coating system.

a) resistance to the environment,
b) resistance to mechanical damage,
c) resistance to undercut corrosion.

Surface coating technology has advanced to the point where these features can be evaluated realistically and linked to lifespan predictions. These scientific principles are proven and can be used to evaluate performance of coatings used on off-shore structures, electricity transmission towers, chemical plants or indeed road vehicles.

The PROHESION™ concept developed by B.P. Chemicals coating subsidiary Mebon Ltd, pre-supposes that "PROtection is adHESION" and research work over the last decade has clearly shown this to be the key! Techniques aimed at the control of adhesion in the long term are therefore critical to the formulation of a high performance system. We can summarise four fundamental principles:

a) the system must have good surface wetting properties,
b) primer should contain electrochemical inhibitors,
c) the system should inhibit sulphur dioxide and chloride ion diffusion,
d) the system should be stress relieved and non-ageing (does not change chemically or oxidise), figure 2.

Surface wetting is essential to ensure that the polymer has the maximum adhesion to the substrate. All coatings are to some extent water permeable. Given that moisture will eventually get to the substrate, corrosion inhibitors are required which suppress electron removal from the steel (such inhibitors would *not* include zinc phosphate, which is a barrier pigment). The movement of vapour in and out of the coating also needs to be controlled to prevent blistering and loss of intercoat adhesion. Vehicle underbodies normally experience wetting and drying cycles. If a coating is too impermeable, moisture will be trapped and adhesion will be reduced.

Finally, lamellar fillers effectively stress relieve high build coatings to ensure long term intercoat adhesion. This is necessary because of physical changes in different layers of a coating system such as temperature, impact absorption or flexing.

## 4 NEW DEVELOPMENTS IN INHIBITIVE PIGMENTS

B.P. Chemicals has recently developed a unique non-toxic anti-corrosive system based on ion-exchange technology. Ion-exchange pigments are produced by ion-exchanging anti-corrosive anions and cations onto porous inorganic oxides, such as silicas. Aggressive ions permeating the paint film are preferentially exchanged onto the inorganic support and the anti-corrosive species released to protect the metal surface. Figure 3 is a schematic

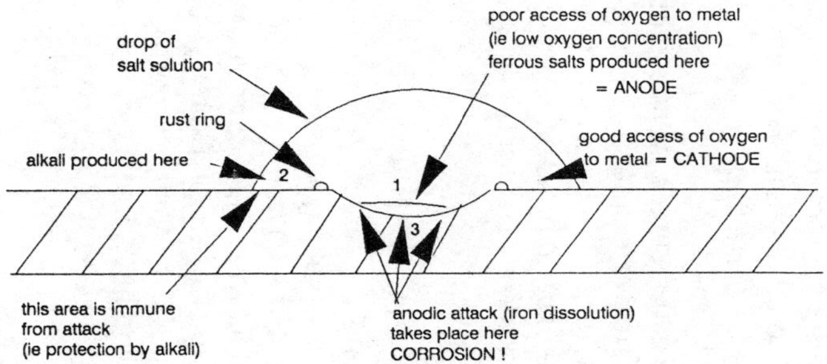


FLOW OF ELECTRICITY WITHIN THE SALT DROP

1) positive ions flow away from point 1 (CATIONS)

2) negative ions flow away from point 2 (ANIONS)

FLOW OF ELECTRICITY THROUGH THE METAL

3) electrons flow away from point 3 toward point 2 (this is the flow of electric current within the metal)

FIGURE 1 the action of a drop of salt solution on iron or steel (illustrating the electrochemical nature of corrosion)

| Performance Requirement | High Performance Coating Features |
| --- | --- |
| Resistance to environment | UV stability; low chloride and sulphate permeability; balanced water vapor permeability |
| Resistance to mechanical damage | Impact strength; abrasion resistance; adhesion |
| Resistance to undercut corrosion | Electrochemical inhibition; adhesion |

FIGURE 2 design life protection

representation of the calcium-exchanged silica mechanism. The advantages of this mechanism are as follows:

a) the anti-corrosive species is used only where and when required, a much more efficient system than exists in traditional pigments.
b) the pigment particles are insoluble and, as the exchange takes place at a molecular level, no change in pigment morphology occurs as the paint film weathers. This reduces void formation and film breakdown, leading to greater protection and longer life.

An extensive series of accelerated corrosion tests, outdoor weathering tests and field trials at various British Petroleum sites have been carried out. These tests showed that primers incorporating calcium-exchanged silica, provide protection superior to that of zinc phosphate, red lead oxide and zinc chromate. This improvement in performance has been demonstrated across a wide range of resin systems.

## 5 THE SCIENTIFIC APPROACH TO VEHICLE PROTECTION

The principles of corrosion prevention described above have been used successfully to protect vehicles for nearly ten years. The system offered by Mebon Ltd is COACHGUARD™ and has the key components illustrated in figure 4. The corrosive environment and also the coating requirements vary depending on the part of the coach or bus to be protected:

Above the Floor Line. Here corrosion is not such a severe problem as steelwork is not exposed continuously to corrosive elements. However, the coating system must have excellent adhesion to the steel substrate to enhance the performance of adhesives which may be used to bond panels or glass on to the vehicle frame and also prevent the spead of undercut corrosion. A single coat of Coachguard Anti-Corrosive Primer 4-1-28 at 35 to 50 microns dry film thickness fully meets the requirements.

Below the Floor Line. In this area the environment is very corrosive with the vehicle underbody certainly being exposed to wetting and drying from water washed up from the road and also in cooler climates being exposed to salts which are used to keep roads free from ice, figure1. Optimum protection is achieved in this severe environment by application of Coachguard Anti-Corrosive Primer to a dry film thickness of 35 to 50 microns followed by a single coat of Coachguard Protective Coating applied at 75 to 125 microns dry film thickness. For coaches and buses exposed to a high level of abrasion from stone chips where roads may be of poor quality a further coat of Coachguard

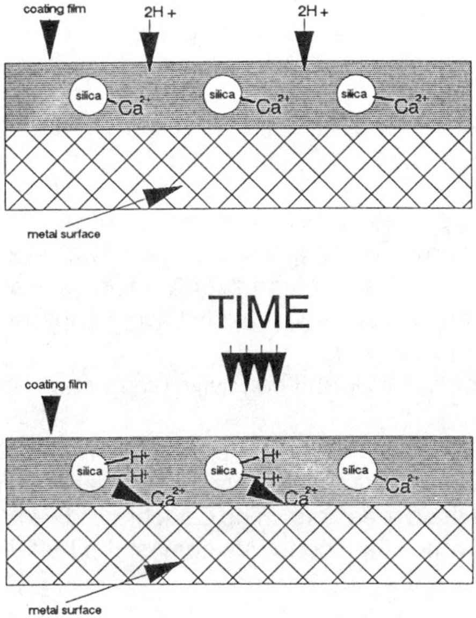


FIGURE 3 the mechanism of ion-exchange corrosion inhibitors

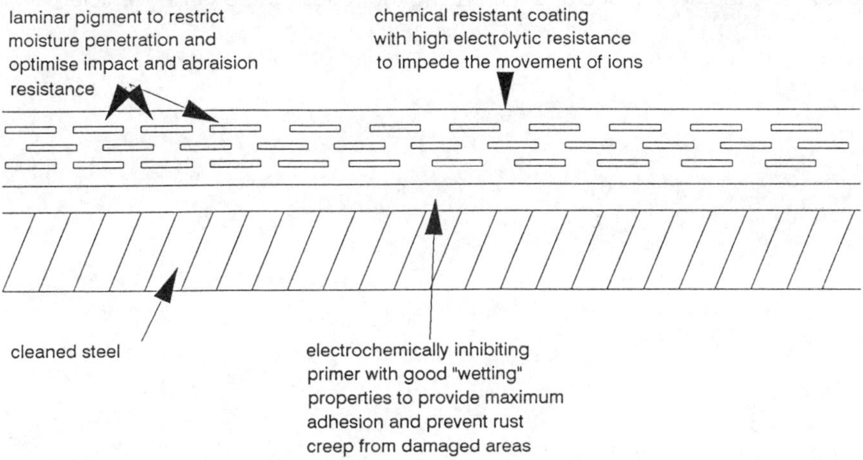


FIGURE 4

Underbody Sealer gives added protection to areas where damage may be expected.

## 6 SUMMARY

The COACHGUARD™ system can be summarised as offering the following productivity benefits:

a) a clean fast drying process which improves productivity,
b) reduced wastage and added flexibility associated with single pack materials,
c) products resist mechanical damage shortly after application meaning that work can be carried out on the vehicle following application of the product without damaging the system,
d) products do not contain toxic anti-corrosive pigments which could constitute a health hazard,
e) Coachguard Anti-Corrosive Primer is a buff yellow colour which contrasts against the grey colour of the steel. this makes materials easy to apply and also inspect for missed areas etc. The Coachguard Protective Coating is a silver grey colour which then contrasts against the buff yellow colour of the Primer.

The application process for these products can be tailored to suit the coach or bus manufacturer with the added bonus of *water based* and *weldable products* both being available. Coachguard will often enhance production methods resulting in a faster throughput of vehicles and cost savings as well as extending the service life and increasing the resale value of the vehicle.

# TOWARDS LONGEVITY OF CRANKCASE LUBRICANTS: CURRENT SITUATION

K. Dickens

Esso Research Centre
Milton Hill
Abingdon
Oxfordshire, OX13 6AE
United Kingdom

## 1 INTRODUCTION

The basic purpose of a lubricant is to minimise the friction and hence wear between two surfaces in motion. The first, obvious, property of a lubricant is therefore its ability to keep moving surfaces apart, in all conditions of pressure and temperature and in the presence of contaminants. Secondly, the lubricant is also required to cool the internal parts of the engine by carrying away the heat generated at the bearing surfaces or reaching them from an outside source such as the heat of combustion. Thirdly, a lubricant must be stable enough to be able to continue to carry out its function for the design lifetime (oil drain interval), even in the presence of outside contaminants such as water and acidic combustion products. The lubricant should also protect the surfaces with which it comes into contact from corrosion as well as suspending water and debris in solution and improve sealing.

Within the last sixty years considerable changes have occurred in engine design and so the demands placed on the crankcase lubricant have also increased. Table 1 summarises the design features which have changed during the last sixty years and table 2 summarises the impact of these on the engine oil in terms of sump temperature and load and also in terms of increased oil drain interval to meet customer requirements.

As the engine designs and customer requirements have changed from 1930 to 1990 the engine oil has been put under more and more stress and the technology used to meet these requirements has changed. The use of additives and changes in base oil quality have meant that it has been possible to meet these evolving needs.

In addition to the changes observed over the last sixty years, future developments are also under way and will continue to push the engine oil to the limit. These trends will include the following:

a) higher quality,
b) longer warranty,
c) better performance,
d) reduced weight and less drag,
e) reduced emissions,
f) advanced engine and transmission designs,
g) lower oil and fuel consumption,
h) longer oil change/service intervals,
i) use of lower viscosity oils, engine 10W-30 and 10W-40, transmission 75W-80 and 75W-90.

## 2 ENGINE OIL SPECIFICATIONS

In order to ensure that an engine oil has the correct properties to lubricate an engine as well as meeting future trends, the engine oil's performance is defined by a series of engine oil specifications. These engine oil performance standards or specifications are set by either national bodies (API, American Petroleum Institute), international bodies (SAE, Society of Automotive Engineers, CCMC, Comité des Constructeurs d'Automobiles du Marche Commun and ILSAC, International Lubricant Standardisation and Approval Committee) or vehicle manufacturers (OEM, Original Equipment Manufacturers) such as Mercedes-Benz, Volkswagen/Audi/Seat, PSA (Peugeot/Citroen/Talbot), Volvo, MAN etc. These specifications are constantly being up-dated owing to design changes and customer requirements. The performance specifications range from the SAE crankcase oil specifications, which relate only to viscosity and related physical properties, to very detailed API specifications for engine oils.

The specifications include physical properties such as viscosity, volatility and concentrations of metals and also performance standards. The performance standards are set by selected combinations of standard laboratory and engine tests together with appropriate limits set on critical parameters such as wear and engine deposits. Tests on lubricants must be carefully standardised because few of the tests measure fundamental properties that are

Table 1 Engine design changes between 1930 and 1990.

| Engine design change | 1930's | 1990's |
|---|---|---|
| compression ratio | 5 | 10+ |
| crankshaft rpm | 3 000 | 6 000+ |
| air/fuel ratio | rich | lean |
| aspiration | natural | charged |
| engine management | mechanical | electronic |
| power/litre,( kW) | 22 | 44+ |

Table 2 Changes in the environment of an engine oil from 1930 to 1990.

| Environment change of engine oil | 1930's | 1990's |
|---|---|---|
| oil temperature, (C), crankcase | 80 | 160+ |
| " piston rings | 150 | 300+ |
| sump capacity, (litres) | 6 | 4 |
| load, (kW/litre) | 4 | 20+ |
| oil consumption, (litre/100km) | 0.4 | 0.04 |
| oil change interval, (km) | 1 000 | 15 000+ |

Table 3 Typical engine tests required for a premium passenger car engine

| Engine test | Property evaluated |
|---|---|
| sequence IID | rust |
| sequence IIIE | high temperature oxidation and oil thickening |
| sequence VE | low temperature sludge |
| CRC L-38 | bearing corrosion |
| Cat 1-G2 | high temperature diesel deposits |
| M102E | black sludge |
| OM 616 | cam and liner wear |
| VW 1431 | diesel ring sticking and piston cleanliness |
| VW 5106 | cam and tappet wear |
| Peugeot TU-3 | valve train wear |
| Peugeot XN-1 | pitting |
| etc. | |

independent of the method and any subsequent change in the equipment or method will also change the result.

The performance specification for various gasoline and diesel engine oils are outlined in figure 1. Interestingly for some requirements, such as CCMC PD-2, both diesel and gasoline performance requirements are used. A premium passenger car engine oil (PCMO) would not only be expected to meet the gasoline requirements, API SG and CCMC G5, but also the diesel requirements, API CD and CCMC PD-2. Gasoline and diesel engines require different engine oil additive technologies and therefore mixed fleet products, meeting both diesel and gasoline requirements, need a compromise (stressing the engine oil still further).

The demands on the engine oil are ever increasing as the engine oil quality increases to meet the growing demands of modern car engines. Europe is a technically demanding and complex market and this has a significant impact on both development time and cost. This impact is summarised in figure 2, which illustrates the number of engine test hours and of engine tests required to test products over the last thirty years. Many engine tests are required for a premium car engine oil (PCMO). These tests are included in table 3 and evaluate areas such as rust, oxidation control, sludge, diesel deposits, cam wear and pitting. In addition to these tests CCMC and OEM also have elastomer (silicone, viton and nitrile) compatibility tests which all respond to different additive chemistries.

The oil drain intervals recommended for super high performance diesel (SHPD) engine oils are significantly longer than for passenger car engine oils. For this reason field trials are also necessary for super high performance diesel engine oils. The current requirements are summarised in figure 3.

## 3 ENGINE OILS

As discussed previously the primary function of a lubricant is to minimise friction and wear between moving parts. In addition it must cool working parts and should have the ability to prevent or minimise corrosion and wear. It should also possess good resistance to oxidation and thermal degradation. Particularly important in automotive gasoline and diesel engines are the cold-starting properties of the engine oil.

Many of these properties are inherent in the base oils from which the engine oils are manufactured. However, base oils alone are not sufficient to meet the lubricant requirements of the modern highly rated engines discussed in the introduction. Chemicals known as additives are therefore mixed with the base

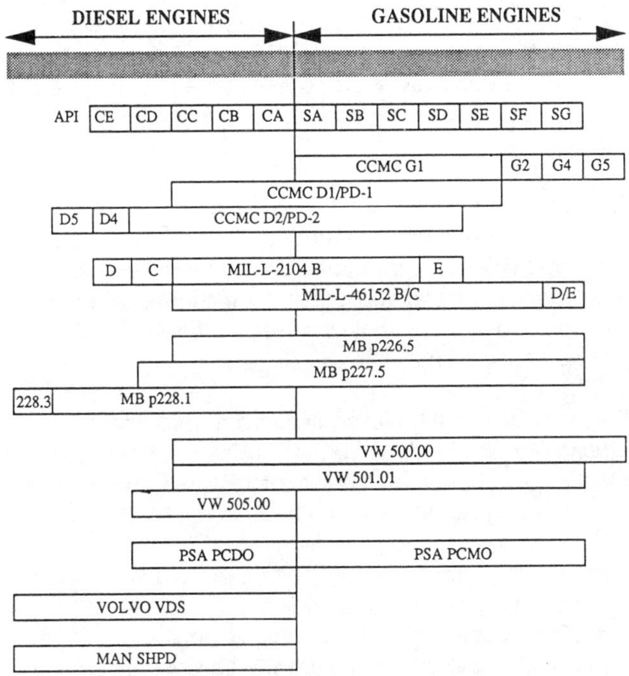


FIGURE 1 : Comparison of Engine Oil Performance Requirements

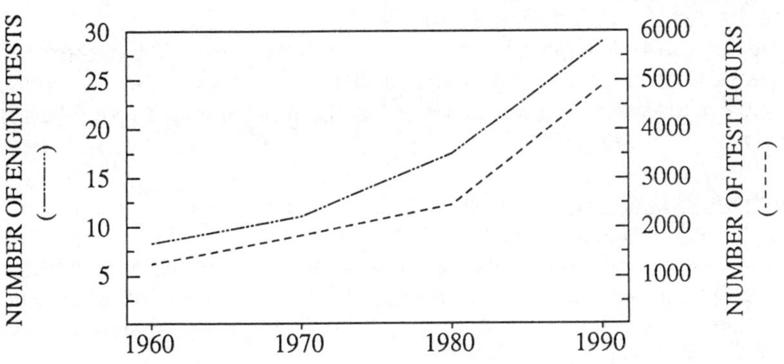


FIGURE 2 : Changes in the Number of Engine Tests and Test Hours from 1960 to 1990

oil to enhance the desirable natural properties, to modify undesirable properties or to impart new properties. Thus a modern engine oil is a balanced formulation made up of carefully selected base oils plus additives such as viscosity index (VI) improvers, pour-point depressants (PPD) and detergent inhibitor (DI) additive package. These are covered in detail in the following sections.

## Basestocks

Basestocks are classified by viscosity and have three key performance areas; a) viscosity and low temperature properties, b) volatility and c) oxidation stability. Low volatility characteristics are essentially inherent in the choice of the base oil for a particular application but the flow characteristic and stability can also be improved by the use of additives. Different basestock quality is needed for different applications.

There are essentially two sources from which the base lubricant fluids, and basestocks are obtained; a) from the refining of petroleum crude oil (mineral basestocks) and b) the synthesis of relatively pure compounds (synthetic basestocks) with properties which are also suitable for lubricant purposes.

Mineral Basestocks. Highly refined petroleum based mineral oils are the main type of basestock used in engine oils. Mineral basestocks are complex mixtures of hydrocarbons generally in the range $C_{20}$ to $C_{40}$ made from the more viscous portion of the crude oil after the lighter fractions have been removed. The principal components of lubricant basestocks are paraffins, cycloparaffins (naphthenes), alkylbenzenes and multi-ring aromatics. The exact ratio will depend on the crude source and the refining process (vacuum distillation, solvent extraction, hydro-finishing and dewaxing). The various components are illustrated in figure 4. The exact basestock performance is therefore determined by its chemical composition (crude source and refinery process) and the various components in the basestock will be subject to different levels of degradation or oxidation.

Synthetic Basestocks. Another source of basestocks is through the synthetic route to produce a higher molecular weight fluid, of fixed properties, by chemical reaction of lower molecular weight material. This is in contrast to refined petroleum oils which are composed of many compounds of varying chemical composition depending on the method of refining and source of crude stock.

Generally, synthetic basestocks have several advantages over mineral basestocks including improved thermal and oxidative stability, more desirable

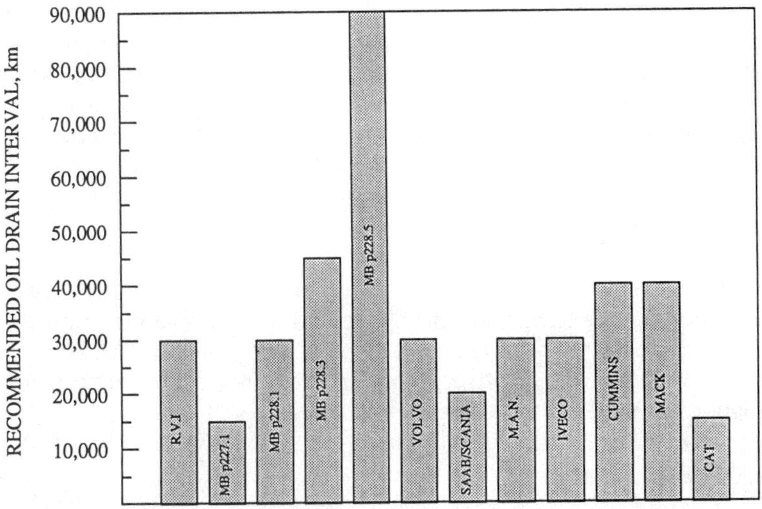


FIGURE 3 : Current Super High Performance Engine Oil Drain Intervals.

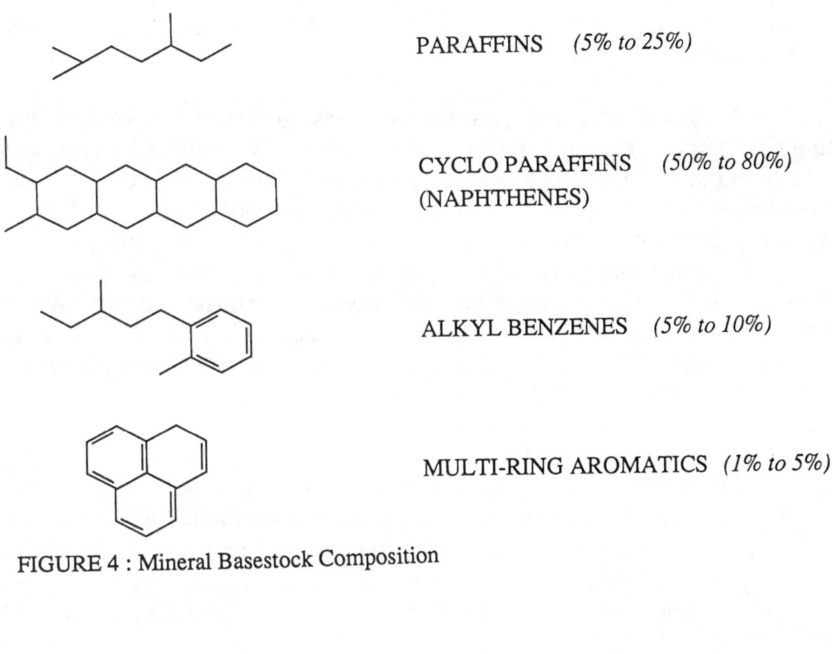


FIGURE 4 : Mineral Basestock Composition

viscosity-temperature characteristics, superior volatility characteristics and preferred frictional properties. Mineral base oils begin to oxidise rapidly at about 120 C but this temperature can be increased to about 150 C with the use of certain additives (anti-oxidants). Synthetic oils, however, do not begin to oxidise until 170 C. One of the principal disadvantages of synthetic base oils, however, is the cost which can be several times more than their mineral equivalent. The use of synthetic basestocks is therefore generally limited to those instances where the performance cannot be met by mineral basestocks alone.

Common synthetic base oils are alkylated aromatics, esters of dibasic acids, *neo*-pentyl polyol esters and olefin oligomers. The most commonly used in synthetic engine oils are PAO (poly-*alpha*-olefins) and polyol and dibasic acid esters. As the requirements placed on the engine oil increase there is a growing trend to higher synthetic contents in engine oils.

## Viscosity Index Improver

The viscosity grade of an engine oil is classified by the SAE (J300) classification (table 4) according to the viscosity at 100 C and at various low temperatures depending on the viscosity grade. High temperature viscosity is related to an oil's consumption and wear characteristics where the low temperature viscosity predicts the cold-starting performance and low-temperature lubrication.

The W grades are based on a maximum low temperature viscosity and a maximum borderline pumping temperature as well as a minimum viscosity at 11 C. Oils without the letter W are based only on the viscosity at 100 C. The low temperature viscosity is determined using a cold cranking simulator (CCS) at high shear rates and has been shown to correlate with engine speeds developed during low temperature cranking. The borderline pumping temperature is measured using a mini-rotary viscometer at low shear rates and is a measure of the ability of an oil to flow to the engine oil pump inlet and provide adequate engine oil pressure during start-up. The oil viscosity at 100 C is measured by the kinematic viscosity and is a useful guide in selecting the correct viscosity grade for use under normal engine operating conditions.

The viscosity of an oil changes with temperature although the changes in oil viscosity are not the same for all oils. At low temperatures the oil is thick and its viscosity is high, but as the temperature increases the viscosity decreases. A measure of the viscosity change with temperature is provided by the viscosity index (VI). Oils with a high viscosity index are usually less sensitive to temperature changes and are therefore better able to perform **efficiently** at

Table 4 SAE (J300) viscosity classification for engine oils

| SAE viscosity grade | Maximum viscosity (cP) | Maximum border-line pumping temperature (C) | Viscosity (cSt) at 100 C | |
|---|---|---|---|---|
| | | | minimum | maximum |
| 0W | 3250 at -30 C | -35 | 3.8 | - |
| 5W | 3500 at -25 C | -30 | 3.8 | - |
| 10W | 3500 at -20 C | -25 | 4.1 | - |
| 15W | 3500 at -15 C | -20 | 5.6 | - |
| 20W | 4500 at -10 C | -15 | 5.6 | - |
| 25W | 6000 at -5 C | -10 | 9.3 | - |
| 20 | - | - | 5.6 | <9.3 |
| 30 | - | - | 9.3 | <12.5 |
| 40 | - | - | 12.5 | <16.3 |
| 50 | - | - | 16.3 | <21.9 |
| 60 | - | - | 21.9 | <26.1 |

both high and low temperatures.

Although the viscosity-temperature quality of a single viscosity grade are often sufficient to meet engine requirements for a particular season or climate, they may be inadequate in a different environment necessitating a change to a different grade. For this reason multi-grade oils are often used. A multi-grade oil is one whose low temperature viscosity and borderline pumping temperature meets the requirements of a W-grade and whose 100 C viscosity meets the requirements on a non W-grade. Multi-grade oils are generally made by the addition of a viscosity index (VI) improver which improves the viscosity-temperature characteristics of conventional basestocks.

A sluggish oil makes engine starting difficult in colder temperatures and may not pump adequately to maintain a satisfactory pressure. On the other hand oils with too low a viscosity can give inadequate lubrication causing wear and high oil consumption. The advantages of a multi-grade oil are illustrated in figure 5. A SAE 10W grade will be thin enough to provide adequate lubrication at low temperatures during start-up, however, it is much too thin at high temperatures and will not provide sufficient lubrication. A SAE 40 grade, however, will give an excellent lubrication at high temperature but will be too thick at low temperatures and may lead to oil starvation. A SAE 10W-40 grade, however, combines the best features of a SAE 10W and SAE 40 grade by behaving as a SAE 10W grade at low temperatures and a SAE 40 grade at high temperatures.

VI-improvers are polymers with molecular weights ranging from 10 000 to 1 000 000. The polymer is swollen by the lubricant in solution and at higher temperatures increases in volume thus having a thickening effect on the engine oil at high temperatures. In addition to the viscosity improvements the performance of these polymers is also dependent on the shear stability or resistance to mechanical shear, where the loss due to shear is reflected in loss of viscosity of the engine oil.

The incorporation of specific chemicals into the chain, such as dispersancy or pour-point depressancy, also allows VI-improver to perform other functions. Various chemistries are used for VI-improvers including olefin copolymer (OCP), polymethacrylates (PMA), hydrogenated styrene butadiene co-polymer (SBC) and hydrogenated polyisoprene. Of these the OCP and PMA VI-improvers (figure 6) are the most commonly used in engine oils.

## Pour-point Depressants

Pour-point depressants (PPDs) prevent the crystallisation of the paraffin

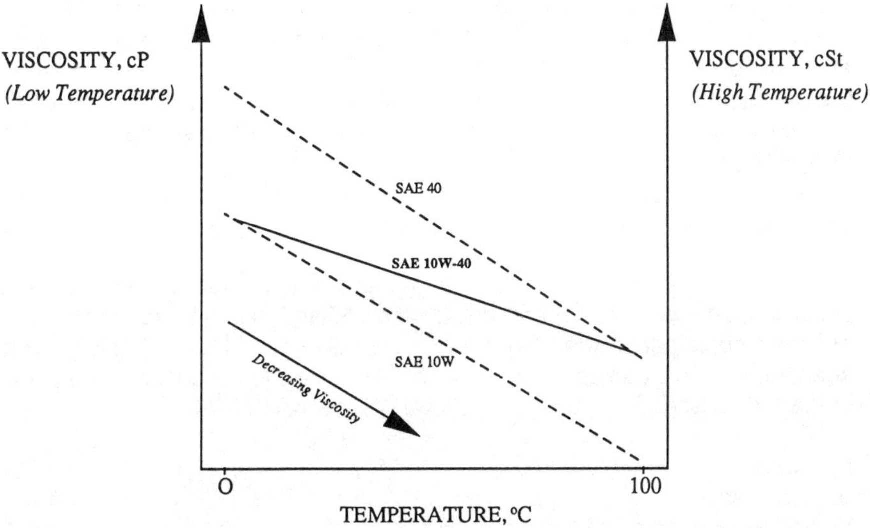


FIGURE 5 : Variation in Viscosity with Temperature

- **OLEFIN COPOLYMER (OCP)**

  MONOMERS

  *ETHYLENE, PROPYLENE, BUTYLENE*

  POLYMER

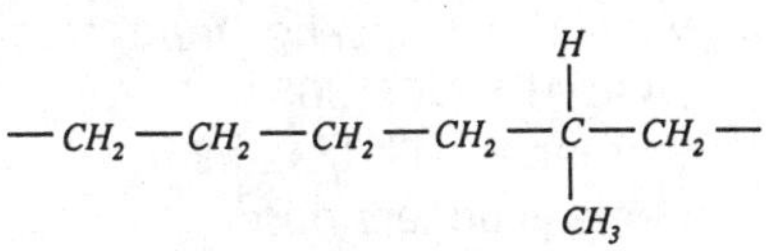

- **POLYMETHACRYLATES (PMA)**

  MONOMERS

  *METHACRYLIC ACID*

  POLYMER

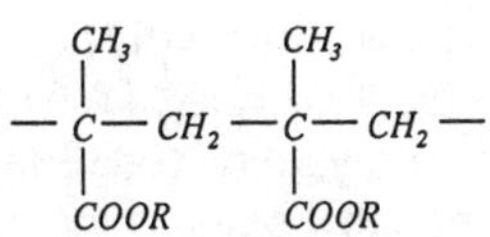

FIGURE 6 : Structure of OCP and PMA VI Improvers

wax which is present in mineral oil. The pour-point depressants do not prevent the wax from crystallising from the oil but are absorbed on to the wax crystals and thus reduce the amount of oil in the crystal thereby reducing the volume of the crystal and permitting flow of the lubricant. Alkylated wax naphthalenes, polymethacylate (PMA) or alkylated phenols are all used as pour-point depressants.

DI Package

A modern engine oil should have good oxidation stability, resistance to corrosion, anti-wear, detergent, dispersant and anti-foamant properties. The additives contained in the DI package generally impart these properties and keep the engine clean and reduce wear. The additives used are listed below and are discussed in detail in the following sections:

a) dispersant,
b) detergent,
c) anti-wear agents,
d) anti-oxidants,
e) corrosion inhibitors,
f) friction modifier,
g) anti-foamants.

Engine oils which met API SA requirements were basically mineral oils. However, as engine designs have changed and additional engine tests have been incorporated into the engine oil specifications the use of additives have become more important. This is reflected in the trends illustrated in figure 7 where the DI package treat increases as the API specification changes from API SA to API SG (1960 to 1990). Interestingly, the treat rate for any specification decreases with time as the additive technology required to meet the requirements is understood.

Dispersant. The dispersant is considered to be the corner stone of any formulation. The dispersant suspends soot, markedly reduces the size of sludge particles, minimises varnish deposits and neutralises combustion acids. Dispersants are particularly important in low temperature gasoline operation to overcome "black sludge". A typical example of the chemistry is poly*iso*butene succinic acid polyamide (PIBSA), figure 8. Typically dispersants contain a polar head attached to a solubilising high molecular weight hydrocarbon chain. The polar head attracts debris and generally contains nitrogen, oxygen or phosphorus. Dispersant treats, however, are generally limited as the nitrogen in the polar head is particularly aggressive to seals and the dispersants also have poor low

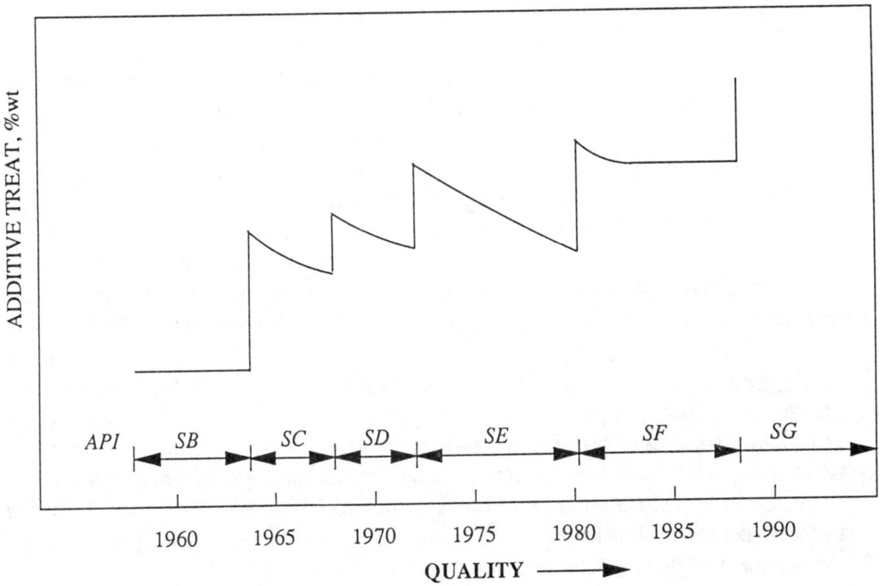


FIGURE 7 : Changes in Additive Treats with API Speicifications

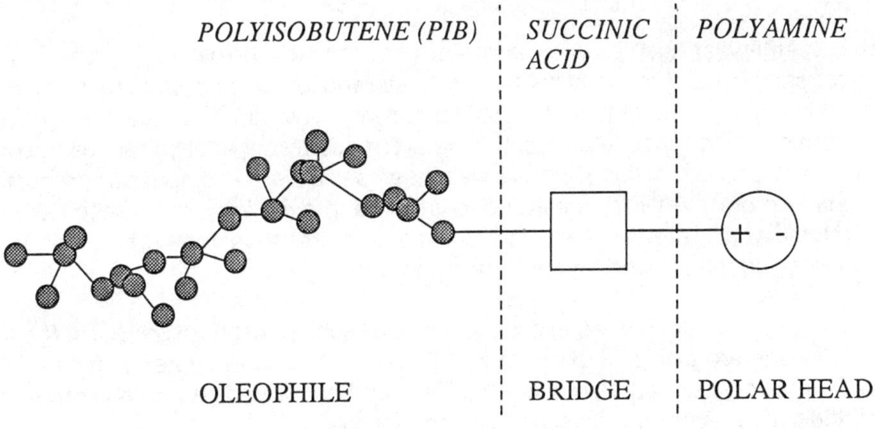


FIGURE 8 : Structure of PIBSA Dispersant

temperature properties necessitating the use of expensive synthetic seals.

One engine test where dispersancy is particularly important is the CEC M102E engine test. The use of a well formulated additive package improves the oil's sludge control and hence its life. This is illustrated in figure 9 where the average sludge is plotted against the test hours in the CEC M102E engine test where a rating of 10 is clean. Engine oils meeting API SA are basically base oils and give poor sludge control and a short drain interval of 1000 km. However, as the quality of engine oils have increased from API SA to the current level of API SG the average sludge merit shows an increase as does the recommended oil drain interval. Again this is representative of the changes from 1960 to 1990.

Detergent. The detergent functions in a similar way to the dispersant having a polar head and solubilising hydrocarbon chain. Dispersants, however, are neutral whereas detergents are basic and as such can neutralise acidic "blow-by" gases from oxidation of the fuel, in particular high sulphur diesel fuel. Detergents therefore also control rust, reduce laquer, carbon and varnish deposits on the piston and the high levels of soap in many detergents prevent ring sticking under severe high temperature conditions.

The function of a typical detergent is illustrated in figure 10 where the detergent contains a metal carbonate (K, Na, Mg, Ca, Li, and Ba) and a surfactant. The hydrocarbon chains can contain either sulphonate, phenate or salicylate and the detergents can be neutral, $M(RSO_3)_2$, basic $M(RSO_3)(OH)$ or over based, $M(RSO_3)(OH) + MCO_3$.

Anti-wear Agents. Wear is loss of metal from engine parts and is generally caused by metal-to-metal contact and if allowed to continue will lead to engine malfunction. Metal-to-metal contact can be avoided by the addition of film forming compounds to protect the metal surface by either physical absorption or by chemical reaction. The anti-wear agents and their decomposition products form a protective film on metal surfaces thereby preventing wear. These reduce engine wear in areas of marginal lubrication such as the valve train, bearings and liners and the top and bottom piston rings.

Typical anti-wear agents are zinc dialkyl dithiophosphate (ZnDDP). These additives are also anti-oxidants and their structure is illustrated in figure 11. Corrosive wear can also occur but this can be reduced using anti-corrosion additives which are discussed in a later section.

Anti-oxidants. Hydrocarbon oxidation occurs by radical chain reaction, leading to the production of hydroperoxides. Anti-oxidants prevent the

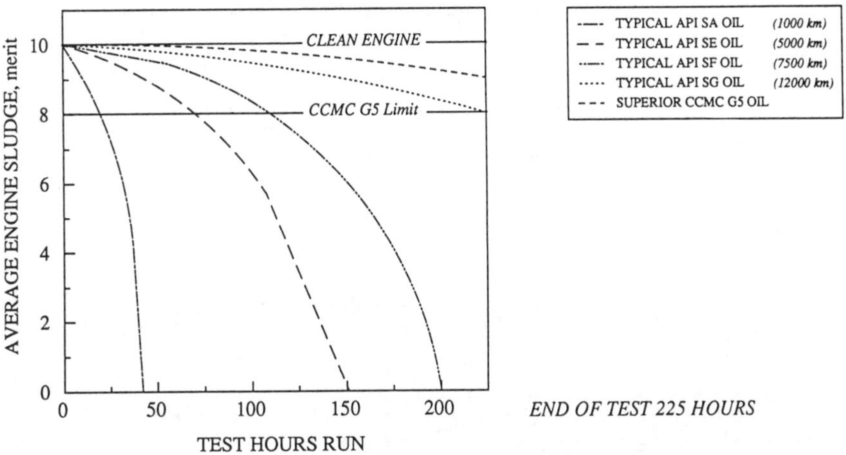


FIGURE 9 : Response of the CEC M102E Engine Test to Engine Oil Technology

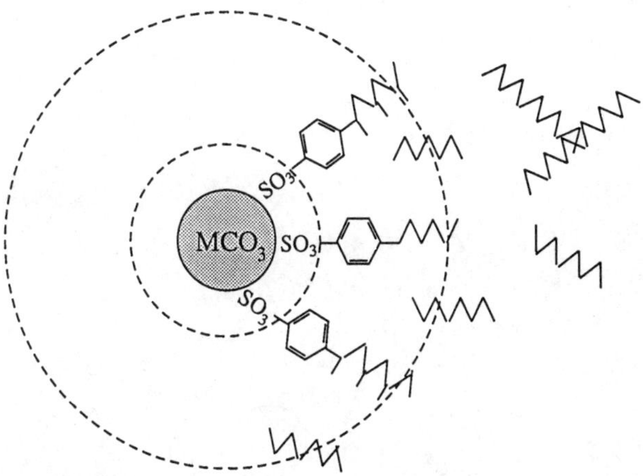


FIGURE 10 : Action of Detergents

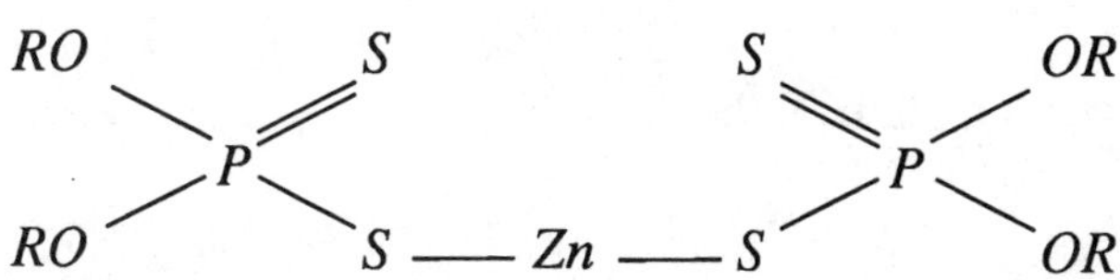


FIGURE 11 : Structure of Zinc Dialkyl Dithiophosphate

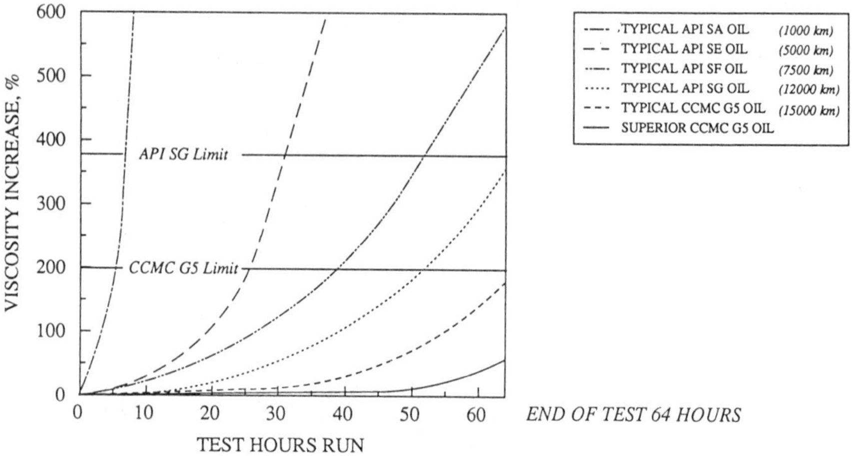


FIGURE 12 : Response of the ASTM Sequence IIIE Engine Test to Engine Oil Technology

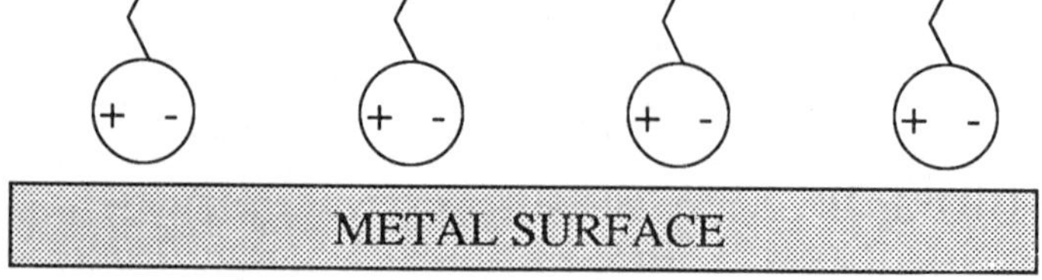


FIGURE 13 : Action of Corrosion Inhibitor

deterioration associated with oxygen attack on the lubricant base oil (viscosity and TAN increase), reduce varnish formation and reduce copper/lead bearing corrosion. These inhibitors function either by destroying free radicals (chain breaking) or by interacting with the peroxides involved in the oxidation mechanisms. Typical compositions are zinc dialkyldithiophosphates, hindered phenols, amines and soluble copper compounds. The level of anti-oxidant must be carefully balanced as these compounds can become pro-oxidants.

The ASTM Sequence 111E engine test evaluates viscosity increase and hence the oxidation control of an engine oil. As with the CEC M102E engine test, the use of a well formulated additive package improves the oxidation control and hence the engine oil's life. This is illustrated in figure 12, where the viscosity increase is plotted against the ASTM Sequence 111E test length. Typically an API SA formulation is a straight base oil and gives poor oxidation control and a short drain interval of 1000 km. As the quality of the engine oil has increased from API SA to the current level of API SG the viscosity increase is reduced and the recommended oil drain intervals therefore increase. Again this is representative of the changes from 1960 to 1990.

Corrosion Inhibitors. The corrosion of bearing material is generally considered to be due to acid reaction with the oxides of the bearing metal. In engines these acids originate from products of incomplete combustion of the fuel which finds its way into the lubricant via the blow-by gases or are produced from oxidation of the lubricant. Anti-oxidants can prevent oxidation and detergents can also reduce bearing corrosion by neutralising the corrosive acids. Other inhibitors not only inhibit oxidation but also form a protective film on the surface making it impervious to attack. Corrosion inhibitors prevent the formation of engine rust and are used as a boost with detergents. The typical components are ethoxylated alcohols and phenols and neutral metal sulphonates. These are illustrated in figure 13.

## 4 SUMMARY

In summary a variety of additives are used to control major engine and engine oil problems. Formulation is the art of combining these additives into an appropriate lubricating oil which both minimises the conflicting effects of various additives and meets the ever changing specifications. Using advanced additive/engine oil technology it has been possible to extend oil drain intervals from 1 000 km to 15 000 km even though operating conditions have become increasingly severe.

# TOWARDS LONGEVITY OF CRANKCASE LUBRICANTS: LUBRICANT ADDITIVES, FUTURE TRENDS

Dr J. Crawford

Adibis
Redhill
Surrey, RH1 1RW
United Kingdom

## 1 INTRODUCTION

The present situation with regard to engine lubricants has been described by Dr Dickens. It was clear from the presentation that 'Longevity' is only one of many aspects of lubricant performance. What should also be very clear is that 'performance' or 'performance specifications' are controlled very closely by the Original Equipment Manufacturers (OEM). Lubricating oil performance is controlled by OEM and changes in engine design affecting the lubricant are their response to technology changes coupled with response to customer requirements and the competitive situation.

An increasing influence on engine performance is Government Legislation either enacted or in the process of enactment. Speaking on behalf of British Petroleum, and perhaps on behalf of a major competitor it is in the interests of companies to market oils which meet the OEM requirements, this of course includes oil drain periods.

## 2 FUTURE ENGINE TRENDS

Changes in engine design invariably result in requests by the OEM to the Oil Companies for lubricant changes. The Oil Companies wish their lubricants to appear in the OEM's vehicle handbook as approved oils. The Additive Companies are aware of the changes and are approached, or approach the Oil

Companies, with the possibility of developing an additive package with the relevant base oil to meet the OEM requirements.

Gasoline Vehicles

The drive for OEM is sales of cars. What this means in simple terms is reliability and/or minimum time in the garage. With respect to the present topic of the lubricating oil the OEM have declared that "roughly" a once a year car service/overhaul is satisfactory for the "average" motorist. There are numerous factors related to the engine which can determine the "life" of the lubricating oil. Important among these factors are the metallurgy of the moving parts, the combustion pressures, the thermal conditions to whch the oil is subjected, the amount of blow-by, the ventilation of the blow-by and the fuel etc. Changes in these factors affect the life of the lube oil.

As we go into the 1990s more emphasis will be placed on fuel economy, hotter running engines and wider multigrades and engine durability. This in turn will result in more stringent engine tests in respect of wear, engine deposit and friction modification. The American OEM requirements will be demonstrated with the development of the VF, IIIF, new sequence VI and alternative to the 1-H2 diesel test if the motor manufacturers hold together.

One major factor (or imposition some would say) is the 'first time' pass philosophy. How this will affect the situation if adopted is open to speculation. At first glance it could be a godsend for Additive Companies. It promises increased additive treatment levels and a boom for the chemical manufacturers....But does it???

In Europe Mercedes and VW in their drive to improve engine design require the oil and additive companies to support them voluntarily or otherwise. Mercedes are proposing a restricted list of approved oils, and the introduction of a new 900 h test on a new engine (M111). This particular test will only be run by Mercedes (but at whose expense?). It is important to note that changes in lubricant requirements in the USA and in Europe are very similar. Generally, it is just a question of degree of severity with respect to the engine performance requirements. The Japanese OEM requirements are similar to the USA and European performance requirements but have a completely different style. Emphasis is on 'genuine' oils with a degree of internal confidentiality between the oil company and the engine builder. Performance requirements are of course similar to those of the leading USA and European OEM. Before long the Japanese Automotive Standards Organisation (JASO) will have established its own set of sequence tests to rival those of the USA and Europe.

Passenger car's lubricants for the 1990s will have to address:

a) high tech engines with multiplugs, multicams, multivalves,
b) tighter emission controls with fuel injection, computer controls, fuel efficiency,
c) new materials in plastics and ceramics.

To add to this a new two-stroke engine design is being investigated by the major OEM. This may revolutionise lubricant formulation.

Summary. It is generally expected that lubricating oils for the future will require improved performance in :- oxidation protection, sludge protection (black sludge), valve train wear control (VTW), and fuel economy. The requirements are more and more improvements to the lubricating oil in terms of engine performance. But the effect of the fuel and the interplay of the fuel on the lubricant should not be overlooked.

Diesel Vehicles

Diesel lubricant issues for the 1990s will be focused around a number of important areas. The major factor of concern is exhaust emission. This tends to dominate the OEM future targets. Requirements for reduced oil consumption will continue with much improved lubrication. The consequences for this will be engine design changes. With this will come new lubricant requirements; Mercedes are already proposing oil drain periods of 90,000 km for top quality oils. In the USA diesel oil quality has been closely linked to the Caterpillar 1G-2 single cylinder engine test for at least 15 years. A new single cylinder engine test to be called the Cat 1K will be introduced during 1990. This will represent all that is new in USA diesel technology with piston configuration of:

a) 3% crevice volume,
b) aluminium,
c) double band for grooves,
d) rectangular second ring,
e) raised keystone top ring,
f) much higher piston temperatures (present 285 to 320 C).

The lubricant requirements for this engine test in terms of additive requirements are as yet unknown. It is suspected that low sulphated ash oils will be preferred. Mack had problems as a result of inadvertently extending oil drain periods. The result was heavy sludge in the oil and oil thickening. When the Cat 1K is in place new oil classifications will be introduced. It is expected that diesel lubricant classifications will be called API CF 2 and API CF 4.

Diesel engine design will place greater emphasis on the lubricating oil. The OEM will be looking at the companies who will be facing demands for improvement in the following areas:

a) improved engine durability,
b) extended drain,
c) low viscosity,
d) improved thermal resistance,
e) fuel efficient lubes,
f) high dispersancy.

It is clear that both gasoline and diesel oils for the future require improvements to the final package which are complementary.

## 3 EFFECTS OF GOVERNMENT LEGISLATION

Broadly speaking, although different counties have different legislation, in general the legislation follows the same lines. The present day issues are those of the "Green" issues. Legislation focuses around emissions. The USA has the addition of Fuel Economy related to the OEM. It is felt that the over-riding concern of the OEM are those issues related to fuel economy and exhaust emissions influenced by legislation. A recent UK government statement indicated that the carbon dioxide emissions from vehicles would stabilise such that in 20 years time they would be the same as at present. If the vehicle population and the distances travelled increase then this will imply more power per unit of hydrocarbon fuel, that is fuel economy combined with more efficient engines. Such engines will of course consume less oil vis longevity!! Shortly after the UK Government statement there were cries from various quarters that this was not acceptable. Carbon dioxide levels emitted from vehicles should be decreased relative to the present situation.

### Gasoline Vehicles

Of the major factors affecting gasoline engine emissions in terms of legislation perhaps the most important are :

Lead content of fuel. Although lead is being progressively removed from fuel, alternatives such as MTBE and alcohols are used as replacements. New octane boost materials are still being sought. The use of these new types of product has given problems in the fuel inlet side of the engine. These have been satisfactorily dealt with by fuel additives. It can be argued that the problems have been moved into the combustion chamber with resulting octane requirement increase (ORI). The interaction between the lube oil and the fuel additives will grow in importance.

Three way catalytic converter (hydrocarbon, nitrogen oxides, carbon monoxide and particulates). The reduction of levels of regulated emissions to low level is the result of the development of successful catalysts. At one stage it was felt that the use of phosphorus containing additives in the lube oil might result in poisoning of the catalyst. Those thoughts are still there, although it must be emphasised that lube oil consumption in modern cars is almost negligible. It is important to note that the catalytic converter increases the amount of carbon dioxide emitted.

Chlorine and bromine levels. With the removal of lead from the oil the amount of chlorine or bromine is substantially reduced. The chlorine and bromine content of lubricating oil can normally be expected to be <1 ppm. The chlorine or bromine content of lubricating oil additives is also low. Depending on the source of the additives, levels in an additive pack can vary between 500 ppm and 20 ppm. However, legislation may aim at a level of <1 ppm. The chlorine content of some additives can be much higher >200 ppm.

### Diesel Vehicles

USA and European emission legislation is almost a topic in itself. It would be inappropriate to cover details of all legislation here. However, it is appropriate to say that such legislation will affect lube oil longevity. This is of particular importance in respect of exhaust gas recycle. In the EEC legislation was introduced in 1988 and will be effective towards the end of 1992. Limits have been set for the pollutants carbon monoxide, hydrocarbons, nitrogen oxides and particulates.

## 4 FUTURE ADDITIVES REQUIREMENTS

From the preceding discussions it is clear that the requirement for additives to help in lubrication are ever more demanding. Speciality chemicals will become even more "specialised". The composition of a lubricating oil can be broken down into the performance chemicals involved.

Lubricating Oil Compositions

A lubricating oil consists of:

a) lubricant base fluid A and lubricant base fluid B,
b) dispersant/inhibitor pack,
c) viscosity index improver.

This finished oil will be expected to meet SAE J300 requirements. Viscometric targets are at 140 C and low temperature (-15 or -20 C etc. via the cold cranking simulator) and borderline pumping temperature must of course meet the OEM engine performance specifications. For Europe there is a Noack volatility test to pass. A similar test in the USA will be brought in shortly.

Lubricant base fluids. There are a number of papers and surveys covering the relative performance characteristics of lubricant base fluids. For the present they will be split into three main categories which are mineral oils, high viscosity index mineral oils or poly-*alpha*-olefins (PAO) and esters or ethers etc. The increasing demand for wider multigrade oils in combination with European OEM Noack requirements makes it very difficult to use mineral oils. Special distillation cuts are required but even then they are generally found to be inadequate. The high viscosity index hydrocarbon oils can achieve the volatility but their lack of polarity leads to additive solubility problems. This is particularly the case with some PAO at low temperature. Oxygenated hydrocarbon products offer all sorts of advantages. Prominent among these are purity, lack of volatility and reduced friction. It is an area of much manufacturing activity at present. Which chemical (solvent) chosen is very dependent on raw material availability and costs. However, it must be stated that failing some wonder star dust to apply to mineral oils then the use of synthetics will increase over the next 5 to 10 years to the detriment of mineral oil. Attempts to define the total market size have been only partly successful. In published information every author covers overlapping areas to study and so a true overall picture is difficult to draw.

Dispersant/inhibitor pack. Poly*iso*-butenyl succinimides (PIBS) are at present the main dispersants used in lubricating oils. Their advantages include thermal stability, superior activity as dispersants, flexibility of modification, availability of raw materials, cost of raw materials and ease of handling. Perhaps their biggest disadvantages in the longer term are the low temperature thickening and the presence of chlorine (50 to 150 ppm). The low temperature thickening is a function of the molecule itself and its inability to rotate because of the methyl groups. If larger quantities of dispersant are required to match the engine test requirements then the problem will be further emphasised.

The chlorine level in PIBS is a function of the method of manufacture; depending on the route it could be 1500 ppm or 50 ppm. Chlorine is a very emotive issue owing to the possibility of dioxan production in the exhaust of vehicles. One PIBS manufacturer has come up with a PIBS which contains no chlorine or bromine. Perhaps this is the PIBS of the future. The design of a new range of dipersants, chlorine free and with no ability to thicken at low temperature combined with the ability to survive at increased engine temperatures in tomorrows engines represents a challenge to chemists the world over. Improvement of the dispersant properties of the existing succinimides has been a long sought after goal of all companies. Do you concentrate on the polar micelle forming end group which is normally an amine? or do you increase the PIB chain and hope that this will give you the properties you require? or both! What other oil soluble polymers capable of "activation" to give dispersant molecules are currently available? At present world manufacturing capability for PIBS is about 500 000 tonnes.

The detergents most commonly used in lubricating oils are calcium based materials. Magnesium and barium products exist but have few advantages over the calcium products when compared with the cost of manufacture. These detergents also contain the ability to neutralise acids produced in the combustion processes and by oxidation of the base fluid. They also clean surfaces and carry dirt into solution. Their improvement probably lies in mixtures of different metals which will somehow create a synergistic effect and lead to remarkable improvement in performance. It would help if the chemical structure of the starting species had inherent antioxidant power. Phenols have always seemed a good starting point from this point of view. There must be others of equal potential. Recent developments in this area have shown new detergents which have exceptional friction modification properties. If ash becomes an issue with the new designs of engine, then such products will be reduced in levels. The use of ashless or low ash base will generally mean the use of nitrogen in some form. The thermal stability of such materials is in some doubt. The world manufacturing capability for these detergents (phenates, sulphonates and salicylates) is about 800 000 tonnes.

As for antiwear additives, zinc dithiophosphates are predominant as they combine antiwear and antioxidant properties. So far it is difficult to see any one chemical species replacing these multi-functional products. If government legislation and OEM limit the use of phosphorus and zinc as seems likely then the search for new performance chemicals is paramount. In the industrial field the use of sulphur compounds to support such performance claims is well known. Prominent among such products are straight sulphurised hydrocarbon products. These would not normally be expected to be able to withstand the thermal rigours of crankcase applications. Among the thermally stable active sulphur compounds currently available are the thiadiazoles. Such specialised products have remarkable properties (and prices to match). Similar products are waiting to be discovered. World manufacturing capability for zinc dithiophosphates is about 200 000 tonnes.

There are many antioxidant species available. This is the area where one feels instinctively that the use of such materials could increase the longevity of the oil to infinity; but there are difficulties. How many of them can stand the thermal rigours of the crankcase environment. When such products break down it generally results in the production of insoluble material and loading on the dispersant. The target operational temperatures are 320 C for the lifetime of the oil. The prominent antioxidants are the hindered phenols (and similar), alkylated diphenylamines, sulphur compounds (sulphides and disulphides) and a number of metal containing products such as those containing copper or molybdenum etc. One aspect of oxidation which has always been a worry is the possibility that the antioxidant may suddenly become a pro-oxidant as the lubricating oil deteriorates. Compounds containing combinations of these chemical species are all well known as is the synergistic combinations of a number of the chemical species (this ignores antioxidant performance associated with other additives). As already mentioned "can they take the forcasted high temperatures?", probably not. A number of these types of product are toxicologically active. Not only are they eye and skin irritant but there is also the possibility that they could be skin sensitisers. The way is open to develop new thermally stable antioxidants or effective combinations of additives to satisfy the new performance targets without running into toxicological problems!

The objective of viscosity index improvers (VII) is to achieve the maximum thickening at a high temperature and the minimum thickening at low temperatures. A further complication to add to this is reduction of viscosity at 150 C under high shear stress. It also helps to use the minimum amount of active polymer. Three types of VII's can be distinguished.

Firstly, non-dispersant VIIs. The thermal breakdown of high molecular weight hydrocarbon polymer creates real problems for the dispersant. This emphasises the importance of using the minimum amount of VII. The big advantage of such products are that they are cost effective. Ploymers are needed which have all the attributes of present VIIs but do not cause this load on the dispersant.

The neutral type VII carries with it sufficient dispersant activity so as not to cause unnecessary loading of the dispersant in the dispersant inhibitor pack. If

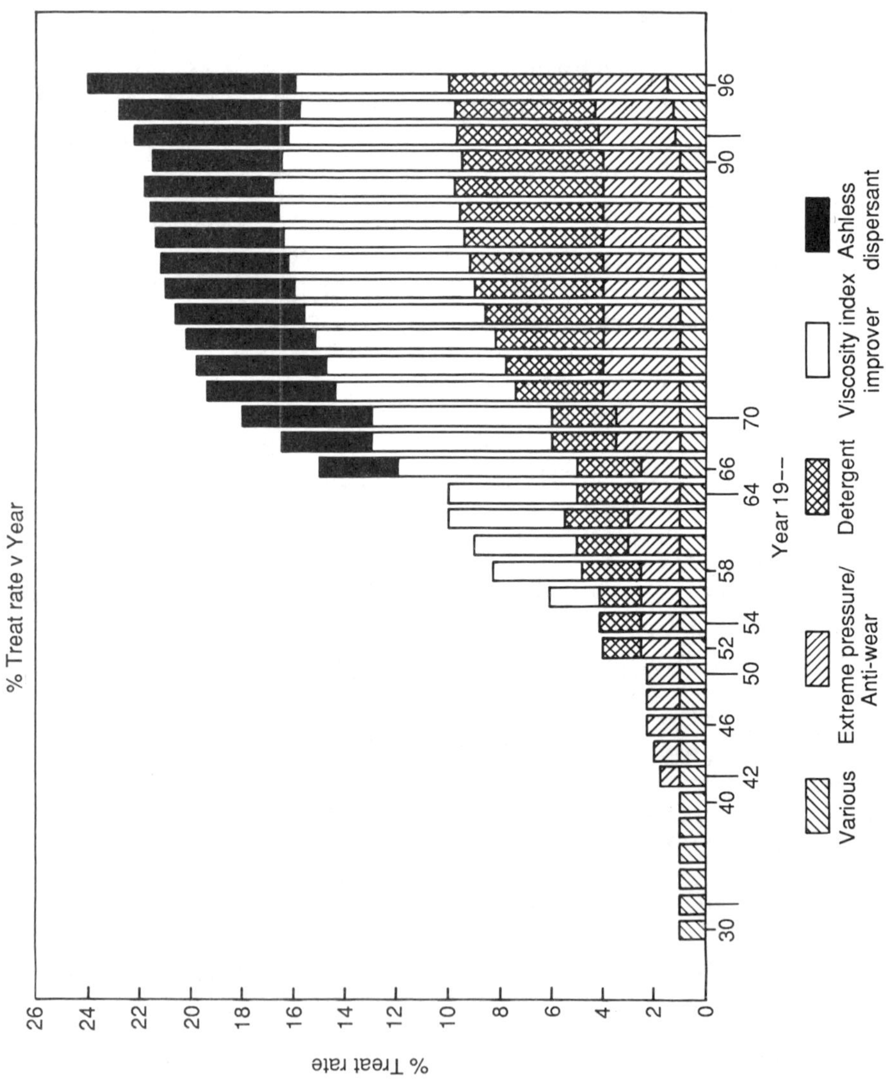


Figure 1 % Additive Content for Top Lube Oils

oil longevity is to be improved the amount of dispersant activity will have to be increased. This will increase the cost of these products relevant to the existing neutral VIIs.

Dispersant VIIs are perhaps the dispersants of the future. Their main disadvantage as seen at present as alternatives to dispersants are two fold. Firstly they take away the opportunity to make monograde oils. This limits the oil companies flexibility to use the dispersant inhibitor pack. Secondly they are expensive relative to standard dispersants. The VII world market is about 0.5 million tonnes of oil diluted material. This represents about 75,000 tonnes of active material.

Multi-functional additives by their nature have some degree of multi-functionality. The use of dispersant characteristics added to thermally unstable VII products is an example of constructive functionality for a lubricant additive. This type of approach taken to develop new additives is one which can lead to some exciting new concepts, which hopefully will lead to commercial products.

Additive Content of Oils. The Additive Technical Committee of Europe (ATC) via their German group (TAD) have produced data on the additive content of oils. Figure 1 illustrates this with an extra view as to how additive treatments might develop in the next five years. If the thermal targets and combustion pressures are increased and lube oil longevity and fuel economy are desired, then this will lead directly to a) improved performance from the dispersant (achieved by either increasing the level of dispersant or the development of more active dispersants), b) improved performance from the antioxidant (achieved either by increasing the level of antioxidant or the development of more active antioxidants) c)improved fuel economy (approached through wider multigrades, friction modifiers and synthetic fluids), the latter can wholly or partially replace the mineral oil this being the most prevalent idea in today's market.

## 5 CONCLUSIONS AND SUMMARY

Government legislation with respect to the environment has a large effect on engine and lubricant developments.

Future engine designs will present new challenges for lubricant and additive chemistry.

To be specific in respect of future additives there is a demand for new dispersant additives, extreme pressure/antiwear additives, detergent additives, viscosity index improvers and antioxidants.

The influence of fuel and fuel additives could be crucial.

# GASOLINE VOLATILITY

A.P. Mercer
Thornton Research Centre
Chester, CH1 3SH
Cheshire
United Kingdom

## 1 INTRODUCTION

The motorist who stops at a filling station on a hot day will often smell petrol as he fills his car. On resuming his journey he may notice that the car starts grudgingly and accelerates poorly during the first mile or two. However, the sun is shining and, like the petrol, his irritation soon evaporates. The same motorist six months later finds his car covered with ice in the morning. He may smell petrol as he starts it, but more important, he may again notice its unwillingness to start and its sluggish acceleration. Because the weather is cold he will probably dismiss this as being normal and therefore acceptable, and give it no more thought until the next time it happens. In either case the motorist is unlikely to think of the fuel as the culprit. He is far more likely to blame the car as a whole, or, more specifically, the battery. He may get as far as dirty spark plugs. In many cases he may well be right, but the quality of the fuel is an essential ingredient in both cases. This paper explains why.

## 2 VOLATILITY AND FUEL QUALITY

Fuel quality is controlled ultimately by the customer, since clearly fuel that is unacceptable will not sell. To help ensure that modern gasoline meets the consumer's requirements, its quality is controlled in three broad technical areas, viz:

a) Volatility: the ability of the fuel to vaporise and mix with air to ensure combustion.

b) Octane quality: a measure of the 'smoothness' of combustion of the air/fuel mixture.
c) Additives: a means of enhancing existing fuel properties or of adding new ones.

The customer's perceptions influence these quality features for the following reasons:

a) Volatility controls *driveability*, the smoothness of response of the vehicle to movements of the throttle. This is not simply a matter of engine size, nor of torque characteristics but a measure of how well the vehicle prepares the mixture of air and fuel. It is particularly important under extremes of cold and hot conditions. Starting, acceleration, idle and warm-up time all contribute to driveability performance.

b) Octane quality controls the occurrence of *knocking* or *pinking*, and, ultimately, of damage to the piston and combustion chamber.

c) Additives improve various aspects of fuel quality. They may for example control deposit formation (detergents), inhibit deterioration during storage (anti-oxidants), reduce carburettor icing, or enhance combustion (spark aiders).

## 3 VOLATILITY AND VEHICLE PERFORMANCE

### Volatility Principles

Volatility can be described as the tendency of a liquid to vaporise. It therefore depends on both the composition of the liquid and its temperature. Composition is important here because gasoline is not a simple liquid; it is a mixture of hydrocarbons (perhaps as many as 300 in a single fuel) that boil at temperatures between about 20 and 220 C. Temperature is important because fuel in a vehicle sees not only the ambient temperature, but also the operating temperatures of fuel system components; overall it may see a range from -20 to +120 C in typical European conditions.

### The Distillation Curve

Gasoline volatility is usually presented as a distillation curve obtained from the ASTM standard test,[1] which measures the percentage of the fuel evaporated at a given temperature. The curve thus describes how the various fuel components vaporise and boil during refining, and dictates largely how the fuel as a whole behaves during service in a vehicle. A typical example is shown in figure 1, with the overlapping temperature scales characteristic of ambient conditions, vehicle fuel system components, and fuel boiling shown together on the vertical axis.

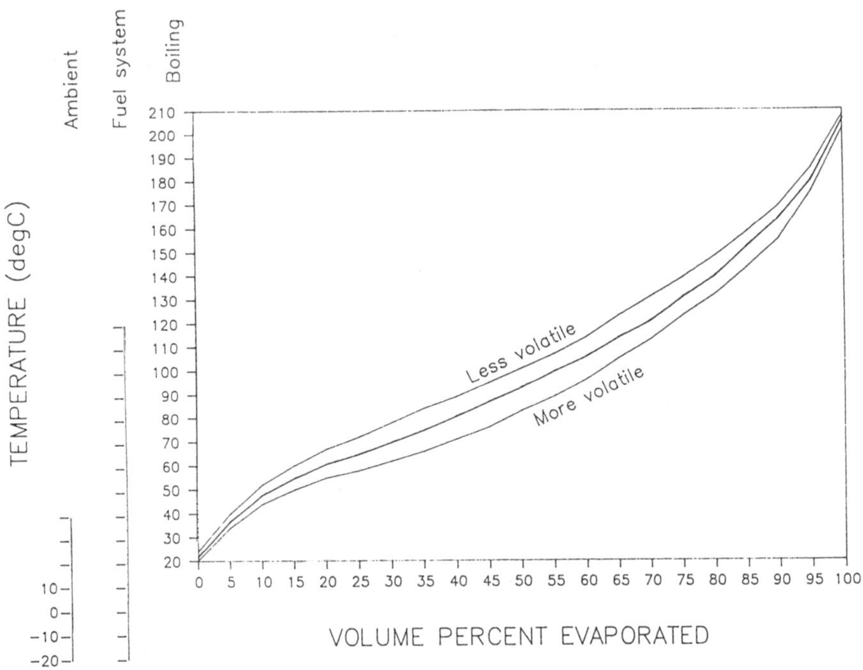


Figure 1 GASOLINE DISTILLATION CURVE
Typical results from an ASTM distillation test

The exact form of the curve is controlled by specifications that are set variously at three different levels; national, within the industry, or within individual companies. These specifications determine not only the absolute levels of fuel volatility, but also the upper and lower tolerance limits at specific points along the curve which have been found to correlate well with particular aspects of vehicle performance. Probably the most informative way to discuss volatility is to follow the distillation curve from the bottom to the top, noting the important features at each stage. We shall begin with the lightest or most volatile components, those at the bottom or 'front end'.

Front-end Volatility. This is conventially measured and controlled via two different fuel parameters. The first is the Reid vapour pressure (RVP);[2] this depends strongly on the amount of the most volatile component in gasoline, usually butane, which boils at around 0 C but dissolves well in gasoline and so in practice is released over a much wider range of temperatures. The second measurement is E70, the percentage evaporated at a temperature of 70 C.

The front-end components in a gasoline influence vehicle performance at both ends of the ambient temperature scale. In cold weather, the fuel strikes cold vehicle components and so may not vaporise sufficiently to provide an adequate air/fuel mixture. Both RVP and E70 must therefore be high enough to ensure satisfactory cold starting. At the other extreme, during hot weather or when the vehicle is re-started after a short break, excessive vaporisation can lead to the formation of pockets of vapour in the fuel system before the fuel reaches the engine. This phenomenon, known as 'vapour lock', leads to an intermittently weak mixture in the combustion chamber and can cause poor hot starting, rough idling, or even stalling in the worst case.[3] Similar effects arise from excessive boiling of the fuel in the carburetter float bowl, a phenomenon known as 'foaming', although in this case the mixture is too rich.[4] RVP and E70 must therefore be low enough to minimise excessive vaporisation.

Following a successful start in either hot or cold weather, the vehicle begins to warm up and many of its components pass through the temperature of 70 C at which E70 is measured. This is well above the ambient temperature range, but well below the temperatures of the components in a fully warm engine. A higher proportion of the gasoline can vaporise and for this condition, therefore, we must look at temperatures of around 100 C and above and the middle range of the curve.

Mid-range volatility. How quickly a vehicle warms up after a cold start is usually gauged by the driver by how soon the acceleration performance returns to its normal or fully-warm level. This 'warm-up performance' has been found to correlate well with the value of E100, the percentage of fuel evaporated at a temperature of 100 C. As with the front-end volatility parameters of RVP and E70, E100 has both lower and upper limits determined by practical constraints.

At the lower limit, the mid-range volatility must be high enough to allow the fuel to vaporise as the vehicle's components warm up from a cold start to the fully-warm condition. This is similar to the front-end volatility requirement, but covers a higher range of temperatures. The upper limit is set by a phenomenon known as 'icing', a problem confined largely to carburetted cars operating in cold humid air.[5] As fuel evaporates in the carburettor, it draws in heat from its surroundings. At ambient temperatures just above freezing, moisture condenses and forms ice in the jets and throttle valve body, blocking the flow of fuel and leading to a weak mixture. This can be equally as serious as vapour lock and may cause similar symptoms, such as rough idling and stalling. Unlike vapour lock, however, the cause often disappears when the car is stopped for inspection since the ice melts rapidly, making ready diagnosis of the problem difficult. The most common remedies are heated vehicle air intakes, and anti-icing additives in the gasoline.

Once the engine is fully warm its fuel system may see temperatures of up to 120 C or so, well above ambient but only about half-way up the gasoline distillation curve. To see the importance of the curve at higher temperatures, we need to consider the top or 'back-end' where the less volatile fuel components lie.

Back-end Volatility. The shape of the back end of the gasoline distillation curve is usually described by the fuel parameters E150 and E180, the percentages evaporated at 150 and 180 C, and by the final boiling point. The engine does not normally reach the temperatures necessary to boil these fractions, and so the quality of the air/fuel mixture relies on evaporation of the fuel before its boiling point, and on the ability of the carburettor or fuel injection system to supply the fuel as a fine spray. Consequently, some of the fuel reaches the combustion chamber in liquid form. This is important to the life of an engine and to its fuel economy once it is fully warmed up.

Liquid fuel in the inlet manifold and combustion chamber can lead to two problems that may reduce either the efficiency or the life of the engine. The first is cleanliness, since partially or un-burnt fuel can cause the accumulation of deposits that prevent the optimum flow of air or fuel to the combustion chamber.[6] The second is the dilution of engine crankcase oil by unburnt fuel retained on the cylinder walls. Both of these result from insufficient vaporisation, and so play a large part in determining the lower limits of back-end volatility.

The upper limit is set largely by the need to balance two conflicting requirements of engine economy, which depends strongly on fuel density. Higher density gives better economy during warmed up operation, such as steady motorway cruising, but lower density, corresponding to higher volatility, is better for frequent cold starting, such as short-journey urban driving as explained above.[7]

## What the Customer Sees

The problems that may be seen by the customer can be summarised by relating them to ambient temperatures, as shown in figure 2. The temperature range from -20 to +30 C covers the seasonal variations over most of Europe, and in each band of 10 C there may be one or more common problems, the causes of which are shown along with the effects of changing the fuel volatility. The band

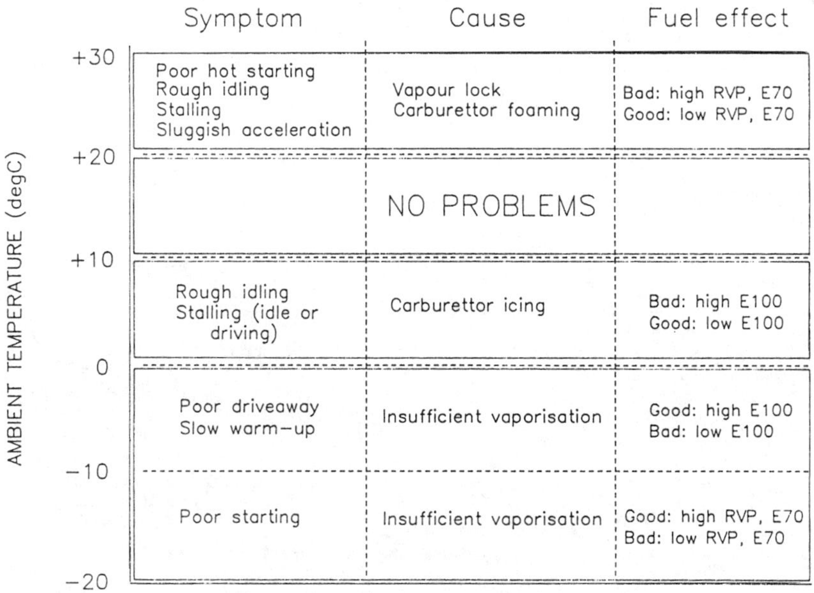


Figure 2 VOLATILITY-RELATED PROBLEMS: WHAT THE CUSTOMER SEES

from +10 to +20 C has historically produced few well-defined problems, and as a result is not covered by the industry standard test procedures for volatility and driveability.[8,9]

This form of summary demonstrates clearly that a careful balance is needed to formulate a gasoline that will meet the various conflicting market requirements. It thus shows some of the constraints that are imposed on gasoline production, and gives a good idea of how these constraints are put together to form fuel specifications.

## 4 SETTING VOLATILITY SPECIFICATIONS

### Conventional Constraints

The effect on gasoline production of the constraints described above is summarised in figure 3, from which it is clear that the different requirements form an envelope in which the distillation curve of a commercial fuel must lie. From this and from figure 2 we can see that no single curve could cover all the volatility requirements of a wide range of climates, either geographical or seasonal, or vehicle technologies, such as those in Europe. In practice, then, the volatility of gasoline is tailored to meet the needs of vehicles in specific markets and particular climatic conditions, down to a level of season-to-season variations in well-defined geographical areas.

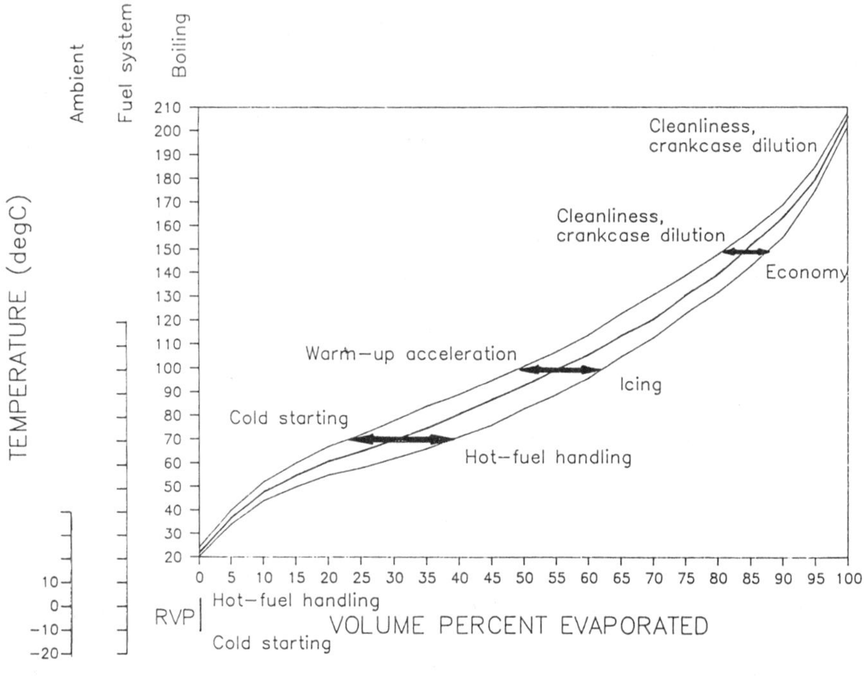


Figure 3 GASOLINE DISTILLATION CURVE
Summary of important properties

For the most part these needs have been derived for a vehicle population dominated by carburettor technology and subject to little environmental concern, and as a result they have changed only slowly over the last ten years or more. Recent changes have been more rapid, however, and this trend is likely to continue for the foreseeable future because of rising pressure in three areas. These are the main constraints on gasoline volatility for the present and the immediate future.

## Current and Future Constraints

Ambient Conditions. Market requirements vary with both climate and seasonal variations. They are rather easier to meet in the extremes of Northern Scandinavia and Southern Italy, or of summer and winter, than in the intermediate levels of mid-Europe or of spring and autumn. Tailoring of volatility levels may be strongly influenced by any move towards geographically wider gasoline specifications, such as that currently being debated for Europe.[10]

Customer and Vehicle Response. Most of the problems described above have been common at various times with carburetted vehicles. However, carburettors are being largely replaced in the market by fuel injection systems that can prepare the air/fuel mixture more precisely. This is having two effects. First, overall vehicle technology is improving, and second, as a result, customer expectations are rising. An additional factor is the trend towards more aerodynamic vehicles, leaving less room in the engine compartment and leading to higher component operating temperatures. The extent to which these two trends will influence volatility requirements remains to be seen.

Legislation. Increasingly strict legislation controls are being imposed, usually led by the state of California and driven mainly by concerns over the environmental effects of vehicle emissions. As a result the recent reductions in gasoline lead content are now being followed by reductions in benzene levels, in evaporative losses, and, via catalytic converters, by partial or total removal from exhaust emissions of carbon monoxide, nitrogen oxides and hydrocarbons. These changes will act directly on what comes out of a vehicle and thus indirectly on what goes into it.

## 5 CONCLUSIONS

a) The control of gasoline volatility is a mature technology. Volatility has been regarded as a bulk fuel property offering little opportunity for commercial advantage, and so has seen little change over the last ten years or more.

b) Volatility specifications have recently begun to receive more attention owing mainly to legislation based on environmental concerns. Growing restrictions on vehicle emissions are likely to be followed by tighter constraints on fuel formulation.

c) Vehicle technology is improving, resulting in more precise preparation of the air/fuel mixture and a rise in customers' expectations of vehicle performance.

d) Continued research and development should ensure that the customer sees few problems, and some benefits, from future changes in vehicle and fuel technologies.

## REFERENCES


1. American Society for Testing and Materials, Standard Test No. D86.
2. American Society for Testing and Materials, Standard Test No. D323, (equivalent to Institute of Petroleum Test No. IP 69).
3. J.M. Jones, J.K. Pearson and J.S. McArragher, "The Setting of European Gasoline Volatility Levels to Control Hot-weather Driveability", SAE paper no. 852118.
4. V.M. Tertois and B.D. Caddock, "Carburettor Foaming and its Influence on the Hot-weather Performance of Motor Vehicles", SAE paper no. 821202.
5. I.C.H. Robinson, "The Effect of Gasoline Additives on Fuel Economy", in "Fuel Economy of the Gasoline Engine", D.R. Blackmore and A. Thomas (eds), MacMillan, 1977, 80
6. B. Bitting, F. Geschwendtner, W. Kohlhepp, M. Kothe, C.J. Testroet and K.H. Ziwica, "Intake Valve Deposits - Fuel Detergency Requirements Re-visited", SAE paper no. 872117.
7. B.D. Caddock, "The Effect of the Physical Properties of Gasoline on Fuel Economy", in "Fuel Economy of the Gasoline Engine", D.R. Blackmore and A. Thomas (eds), MacMillan, 1977, 68.
8. "Cold Weather Driveability Test Procedure", CEC M-08-T-83, Co-ordinating European Council for the Development of Performance Tests for Lubricants and Engine Fuels, 1984.
9. "Hot Weather Driveability Test Procedure", CEC M-09-T-84, Co-ordinating European Council for the Development of Performance Tests for Lubricants and Engine Fuels, 1985.
10. Comitée Européan de Normalisation, Working Group 21, 1990.

# THE ROLE OF ETHERS IN THE FORMULATION OF CLEAN PETROL

M. Dawson,

ARCO Chemical Europe, Inc
Maidenhead
Berkshire, SL6 1YP
United Kingdom

## 1 INTRODUCTION

Despite criticisms to the contrary (especially with respect to carbon dioxide emissions), the USA leads the way in some important aspects of environmental improvement. This is particularly true in the area of vehicular pollution.

The air pollution in places like Denver and Los Angeles have world-wide notoriety. But at least the problem is being addressed and the results are coming through. In Europe, similar pollution exists in many cities, for example Athens, Turin, Milan etc., but no effective strategy for resolving the problem has yet been proposed. It is instructive to examine which of the technical measures being implemented in the USA would be most effective in Europe.

Catalytic converters have been standard equipment in America since 1975 and today only about 5% of the total car population does not have them. These do an excellent job, but with certain limitations. They are only effective when they have achieved a certain operating temperature and they only affect exhaust emissions.

When ambient temperatures are high, more hydrocarbons are released through evaporation of the fuel than through exhaust emissions. To address this, carbon canisters are fitted to the majority of American cars to capture the evaporative losses and then direct them to the combustion chamber. Furthermore, petrol stations are equipped with the means to prevent the release to atmosphere of the saturated vapours which are displaced when vehicles are refuelled.

In spite of having implemented these measures some fifteen years ago, severe urban air pollution still exists in many parts of the USA, and the next phase of controls are now being studied. These will be embodied in the amendments to the Clean Air Act which is now passing through Congress. A key element in the amendment is the requirement that gasoline sold in the worst ozone non-attainment areas should be "reformulated" in order to reduce emissions. The principal provisions of the amendment are summarised in figures 1 and 2.

Most European cities have not yet begun to benefit from catalytic converters and carbon canisters. Even when they are introduced, the USA experience tells us that in due course they will have to do more; catalytic converters and carbon canisters will not solve the air quality problems on their own. What is even more important is that it will be many years before a significant proportion of the European car population is equipped in this way. One of the main advantages of reformulated gasoline is that it reduces emissions from all cars, including the "high emitters" on the roads today.

So, sooner or later, Europe will begin to clean up its gasoline in the same way as has been shown to be necessary in the USA. It is here that the chemical industry has a major role.

## 2 REFORMULATION OF GASOLINE

Gasoline is a mixture of various streams produced from the refining of crude oil. The refiner's objective is to maximise the value of his barrel by utilising his installed units and blending the products from them in such a way as to meet performance specifications at the lowest possible cost. Reformulating goes a step further. Instead of simply aiming to meet the performance specifications, an additional goal of making the product environmentally less damaging is applied.

The componential breakdown of gasolines vary widely depending on the grade and the refinery configurations from which they are sourced. In turn, the composition has a profound effect on the level of emissions from the vehicle. Although there is widespread acceptance that gasolines can be made less environmentally harmful, the details of which species should be maximised and which should be minimised are still being deduced. It is with this goal that the USA oil and automotive industries have embarked on a major, $14 million study to determine the optimum recipe for reducing emissions.

# CLEAN AIR ACT
# S-1630......HR-3030

| PROVISION | | SENATE | HOUSE |
|---|---|---|---|
| CO n/a areas | | 3.1% Oct-Mar from 91 | 2.7% from '92.<br>Poss inc to 3.1%<br>by 2000 |
| Ozone n/a areas | H/C | 15% redn VOC's<br>(during summer) | 15% redn ROG's '92<br>20% '97<br>25% '00<br>Tighter ex specs |
| | NOx | No increase | 1g/gal-->0.4 by'96<br>0.125 if EPA say |
| | O2 | 2% beg.'92<br>2.5% beg.'93<br>2.7% beg.'94 | Same |

Figure 1

| PROVISION | | SENATE | HOUSE |
|---|---|---|---|
| | Aromatics | 30% beg.'92<br>28% beg.'94<br>25% beg.'96 | SAME |
| | Benzene | 1% beg.'92 | 0.8% beg.'92 |
| SCOPE | | 9 cities<br>Others '92 | 9 cities '92<br>Others '94 |
| "Hazardous air pollutants" | | No provision | 15% redn beg.'95<br>20% beg.'97<br>25% beg.'00<br>(cf conventional) |

Figure 2

This study was precipitated by the direction taken by President Bush's Clean Air Act amendments as they passed through the Senate and the House of Representatives. It is now clear that reformulated gasoline will be classed as a clean fuel, and its principal characteristics have been stipulated. These are:

| | |
|---|---|
| maximum level of benzene | 0.8 or 1.0% |
| maximum level of higher aromatics | 25% by 1996 |
| minimum level of oxygen | 2.7% by 1994 |

It is this last requirement that introduces the chemical industry into the picture, for oxygen-containing compounds are not normally produced in the refining process. Increasingly, however, oil companies are embracing the idea of the "open" refinery relying not only on its own production capabilities, but also on products brought in from other industries.

ARCO Petroleum Products Company was the first company to offer a reformulated gasoline in the USA, but its example has already been followed by seven other oil companies, including giants like Shell and Exxon.

The specification of this reformulated gasoline, EC-1 is very close to the specification that subsequently became embodied in the proposed amendments, figure 3. Other reformulated gasolines display many of these same characteristics, but there are still some significant differences, particularly with respect to the levels of benzene and higher aromatics.

## 3 THE ROLE OF MTBE

Both alcohols and ethers can be used as vehicles for introducing oxygen into the gasoline pool. The alcohols, particularly ethanol, display a variety of significant disadvantages relative to the ethers, including high vapour pressure, poor water tolerance and high heat of vaporisation. Hence the most widely used oxygenate is methyl *tertiary*-butyl ether (MTBE). This has been blended into both USA and European gasolines for many years, primarily because it has an extremely high octane value which has enabled it to take the place of the lead compounds gradually being phased out. Last year, some 2 million tonnes were blended into the European pool, and approximately twice that amount into the USA pool.

The focus is now very much on MTBE for its ability not only to add octane, but also to reduce emissions. MTBE is produced on many refineries by reacting *iso*-butylene from the FCC unit with methanol. This is an increasingly economic means for the refiner to add octane to his gasoline pool. Many such units have been started up in the last few years and many more are at present under construction. However, the enormous growth projections for the chemical which are implied by reformulated gasoline programmes will require investment in world-scale plants supplying the merchant market.

There are two basic routes for the production of MTBE. The difference lies in the flow scheme to produce the *iso*-butylene molecule. Thereafter, the reaction with methanol is relatively straight forward and common to both routes. The ARCO process manufactures MTBE as a co-product with propylene oxide, figure 4. ARCO's production represents some 40% of world-wide capacity. Other companies make MTBE as their main product, in which case the route is via the isomerisation and dehydrogenation of field butanes to produce the *iso*-butylene and thereafter reaction with methanol in the usual way. This route is illustrated in figure 5.

## 4 HOW MTBE REDUCES POLLUTION

To understand the chemistry by which MTBE cuts vehicular pollution, it is necessary first to examine which pollutants are of importance. The three most significant pollutants are ground-level ozone, nitrogen oxides (NOX) and carbon monoxide (CO). Their order of importance depends on several factors, especially the climate and topography of the effected area. They are all injurious to human health, especially for individuals with existing cardio or pulmonary conditions, and, in addition, ozone causes damage to agricultural crops.

Ozone is the product of photochemical reactions between hydrocarbons and NOX. Since, in most areas, these reactions are hydrocarbon limited, the mechanism by which to control ozone is the reduction of hydrocarbons going into the atmosphere. Limiting the release of hydrocarbons depends on lowering the vapour pressure of the gasoline in order to reduce evaporative emissions, and also adjusting the composition of the mix to minimise exhaust emissions.

The relative importance of the evaporative and the exhaust emissions depends strongly on the volatility of the gasoline and the ambient temperature, as illustrated in figure 6. This shows that at the temperatures prevailing in those areas which have ozone episodes, the evaporates dominate the total emissions of hydrocarbon released.

### Evaporative Emissions

In terms of evaporative emissions, MTBE has a very direct role. The significance of a particular species depends on its photochemical reactivity, that is, its tendency to react with NOX in the presence of sunlight to form ozone. Also

## Arco Products : EC-1
## Emission Control Gasoline

Unleaded "Leaded Regular" Gasoline

Spec :

| | |
|---|---|
| Lead | none added |
| Summer RVP | 8 psi max |
| Benzene | 1% max |
| Aromatics | 20% max |
| Bromine no. | 20 max |
| Sulphur | 300 ppm max |
| Oxygen | 1.0 % max |

Major Components:

MTBE
Alkylate
FCC Gasoline
Lt Straight Run

" No " Reformate

Figure 3

## ARCO Route to PO + MTBE

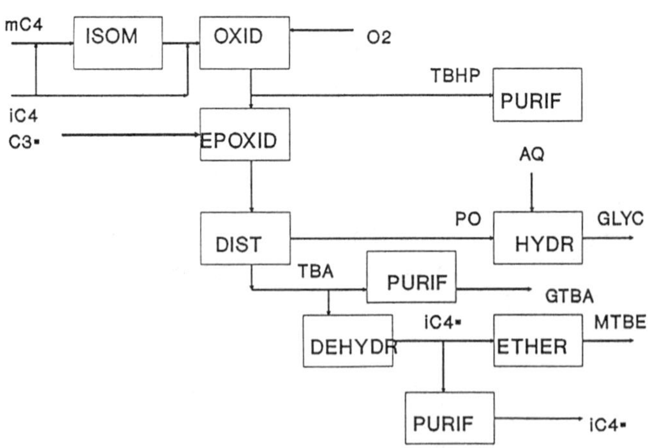


Figure 4

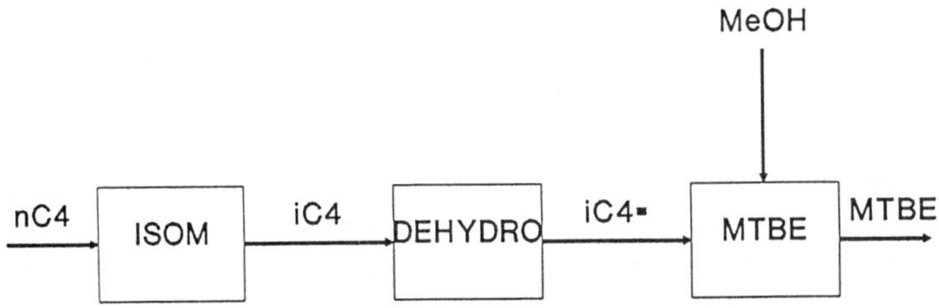


Figure 5 MTBE FROM ISOM DEHYDRO

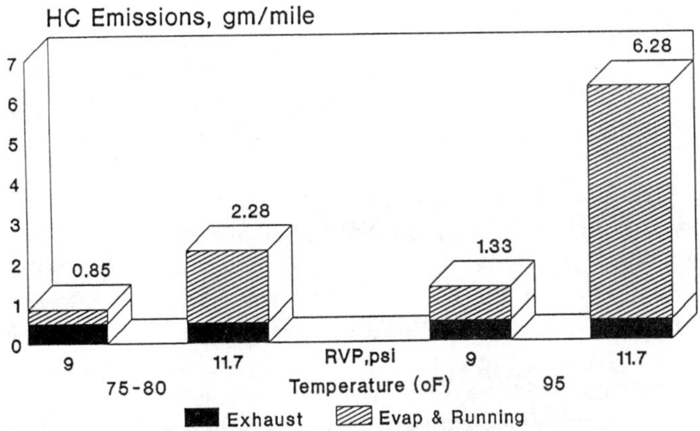


Figure 6 Passenger Car HC Emissions (1981 and newer models)

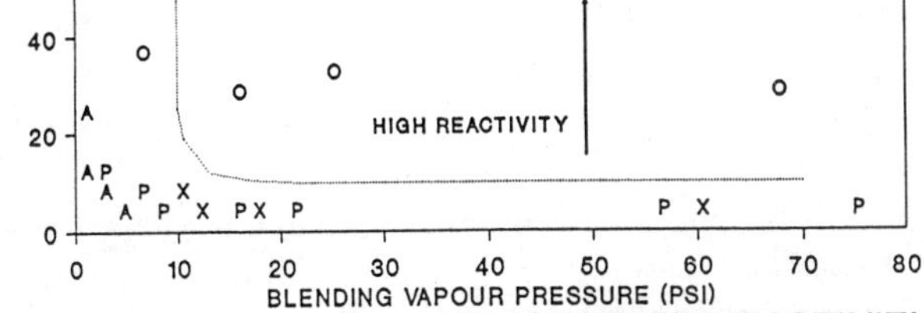


Figure 7 ATMOSPHERIC REACTIVITY OF VOLATILE GASOLINE HYDROCARBONS

important is the compound's vapour pressure, since if it is not sufficiently volatile to escape to the atmosphere, its reactivity is of no particular relevance.

The reactivity and volatility of some typical gasoline components are listed in table 1 and plotted in figure 7. What immediately becomes clear is that it is the light olefins which fall within the envelopes of high reactivity and high volatility. Besides being able to replace their octane contribution if these compounds are removed from the pool, MTBE is also an economic compound into which to convert some of them.

Exhaust Emissions

Exhaust emissions of hydrocarbons are somewhat more complex. Many studies, including the auto/oil work referred to earlier have endeavoured to identify the properties of a gasoline which lead to increased emissions.

The role of aromatics on vehicular pollution has been recognised in the Clean Air Act amendments which stipulate a maximum of 25% by 1996. All the "regulated" pollutants (CO, NOX and hydrocarbons) increase with aromatics content, figure 8, but the data show a considerable degree of scatter, which implies that other parameters are also implicated. Heat of vaporisation appears to be important, since a large proportion of total emissions occur when the engine is cold and there is no source of sensible heat with which to vaporise the fuel. Incomplete combustion then occurs and levels of CO and unburned hydrocarbons in the exhaust increase correspondingly.

Again, MTBE offers an economic solution. The reason aromatic levels have increased over the years is because they are important contributors of octane to the pool. MTBE enables their levels, and the corresponding pollution problems, to be reduced. Similarly, since it has a relatively low heat of vaporisation, MTBE improves fuel mixing and atomisation during cold operation and consequently reduces emissions. Alcohols are particularly poor in this respect, which is one of the problems encountered in trying to design engines to run on pure methanol.

## 5 CONCLUSION

As far as the USA is concerned, reformulated gasolines are seen as being a very important weapon in the effort to clean up urban air since it is now recognised that catalytic converters and carbon canisters are unable to complete the task on their own. In Europe, the same conclusion will undoubtedly be reached in due course, but at this stage there is not yet an appreciation that

Table 1 Atmospheric reactivity of volatile gasoline & tailpipe emission compounds

| Hydrocarbon Type | Compound Name | Atmospheric Reactivity (-OH rate)* | Blending Vapour Pressure (psi) |
|---|---|---|---|
| x | methanol | 1.0 | 60.0 |
| x | *t*- butanol | 1.1 | 10.0 |
| p | propane | 1.2 | 190.0 |
| a | benzene | 1.3 | 3.2 |
| p | *iso*- butane | 2.5 | 75.0 |
| x | MTBE | 2.6 | 8.0 |
| p | *n*- butane | 2.7 | 55.0 |
| x | ethanol | 3.4 | 18.0 |
| p | *iso*- pentane | 3.6 | 21.0 |
| p | *n*- pentane | 5.0 | 16.0 |
| p | *iso*- hexane | 5.0 | 6.8 |
| p | *n*- hexane | 5.6 | 5.0 |
| a | toluene | 6.4 | 1.0 |
| p | *n*- heptane | 7.3 | 1.6 |
| o | ethylene | 8 | 250.0 |
| a | *p*- xylene | 11 | 0.3 |
| x | formaldehyde | 11 | 60.0 |
| o | propylene | 17 | 226.0 |
| a | *m*- xylene | 23 | 0.3 |
| o | pent-1-ene | 30 | 19.5 |
| o | but-1-ene | 30 | 65.0 |
| o | 3-methyl-but-1-ene | 32 | 27.0 |
| o | hex-1-ene | 36 | 6.0 |
| o | *iso*- but-1-ene† | 55 | 66.0 |
| o | but-2-ene | 65 | 49.0 |
| o | pent-2-ene | 68 | 15.5 |
| o | 2-methyl-but-1-ene§ | 70 | 19.0 |
| o | 2-methyl-but-2-ene§ | 85 | 15.0 |

p - paraffins, o - olefins, a - aromatics, x - oxygenates

† - feed for MTBE or *t*- butanol, § - feed for TAME

* - R. Atkinson, *Chem. Rev.*, 1985, **85**, 69 - 201

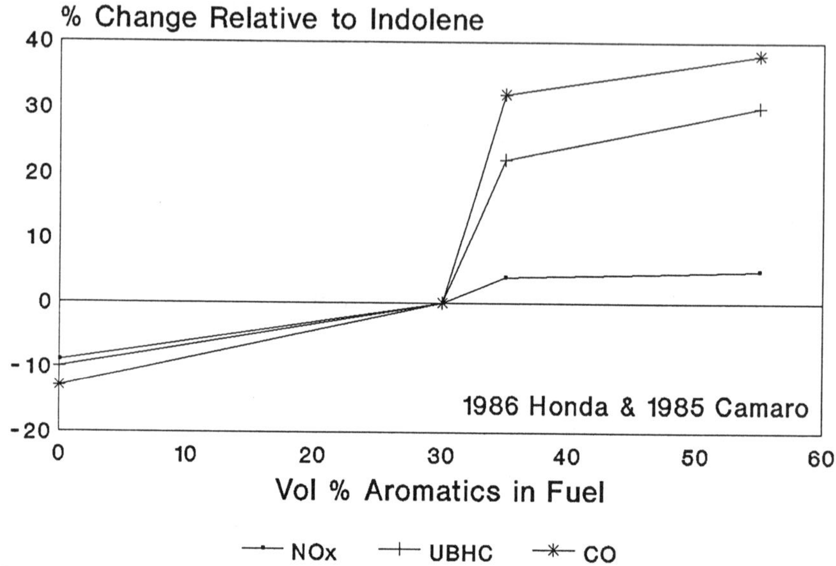


Figure 8 Effect of Aromatics on Exhaust Emissions

- Reduces both exhaust and evaporative emissions
- Reduces all main pollutants
  - CO
  - HC
  - NOx
  - CO2
- Reduces emissions from all cars
  - old
  - new
  - with cats
  - without cats
- Effective immediately
- No changes required to distribution system

Figure 9 Reformulated Gasoline
Summary of Benefits

converters and canisters will not achieve the air quality goals unassisted. Nor is there any strategy for reducing emissions from the existing car population, other than banning traffic from city centres which should surely be considered only as a last resort, not the first line of defence.

MTBE is an integral part of reformulated gasolines. It makes a contribution in a number of different ways, figure 9, and as far as we know, has no significant disadvantages. It can be produced from readily available materials and has been the refiner's octane booster of choice for many years, certainly since long before the legal requirement to include it in the pool. It is for all these reasons that it is now estimated to be the world's fastest growing petrochemical.

# REPLACEMENT OF DIESEL BY ALCOHOL

J.M. Betton

ICI Chemicals & Polymers Ltd
Wilton
Middlesbrough
Cleveland, TS6 8JE
United Kingdom

## 1 INTRODUCTION

Over the past two years, concern for the environment has increased dramatically. One of the main topics of debate has been global warming, but also of concern are high concentrations of particulates, nitrogen oxides (NOX), and ozone in urban areas.

The diesel-fuelled compression ignition (CI) engine is the preferred powerunit for all commercial transport applications because it is intrinsically about 30% more efficient than an equivalent petrol engine. The main drawback of the diesel-fuelled CI engine is the high level of particulate emissions compared with the petrol engine.

The search for the technology to provide the best of both worlds has led to the consideration of alternative fuels for CI engines, the prime contenders being methanol and ethanol. The addition of a few per cent of AVOCET chemical ignition improver allows alcohols to replace diesel as the fuel for CI engines without the need for mechanical spark or glow plug assistance.

## 2 LEGISLATIVE POSITION

In the European Economic Community (EEC), the NOX limit for heavy duty engine emissions is currently set at 14.4 g kwh$^{-1}$ but, as recently as May 1990, the EEC proposed limits of 8.9 g kwh$^{-1}$ NOX and 0.33 g kwh$^{-1}$ particulates to apply from July 1992, with even stricter limits proposed from October 1996. The 1992 limits can just about be achieved by turbocharged (T/C) diesel engines, but not by normally aspirated (N/A) engines, figure 1.

In the USA the Senate Clean Air Bill 1630 has been "reported-out" with weaker provisions than in President Bush's original proposals, but the House Bill 3030 contains alternative fuels proposals deleted from the Senate version. The US Clean Air Act debate is now in a critical, possibly final, stage where the House and Senate, through its selected conferees tries to resolve the differences between the Bills passed by each body.

Regardless of the requirements of the Clean Air Bill, it is likely that action will be taken in the nine urban areas which are classified as "extreme" or "severe" air quality non attainment areas. Federal air quality standards are set in terms of lead, oxides of sulphur (SOX), NOX, particulates, ozone and carbon monoxide (CO). Diesel engine emissions contribute primarily to particulate and NOX levels, to a lesser extent to SOX and CO levels, but not to lead levels. Since ozone is formed by the reaction of NOX with hydrocarbons in the presence of sunlight, diesel emissions also contribute indirectly to ozone levels. In an effort to improve air quality, the Environmental Protection Agency (EPA) has proposed the following limits for diesel emissions.

| Date | Particulates (g bhp$^{-1}$ h$^{-1}$) | | NOX |
|---|---|---|---|
| | trucks | buses | (g bhp$^{-1}$ h$^{-1}$) |
| 1991 | 0.25 (0.34) | 0.10 (0.13) | 5.0 (6.7) |
| 1994 | 0.10 (0.13) | 0.10 (0.13) | 5.0 (6.7) |

These limits represent the current target for the world's diesel engine manufacturers, figure 1.

## 3 DIESEL DEVELOPMENTS

Initial attempts to meet the EPA limits using conventional diesel fuel have been unsuccessful. Recently the use of low sulphur (0.05%) diesel fuel and more sophisticated engines with the addition of:

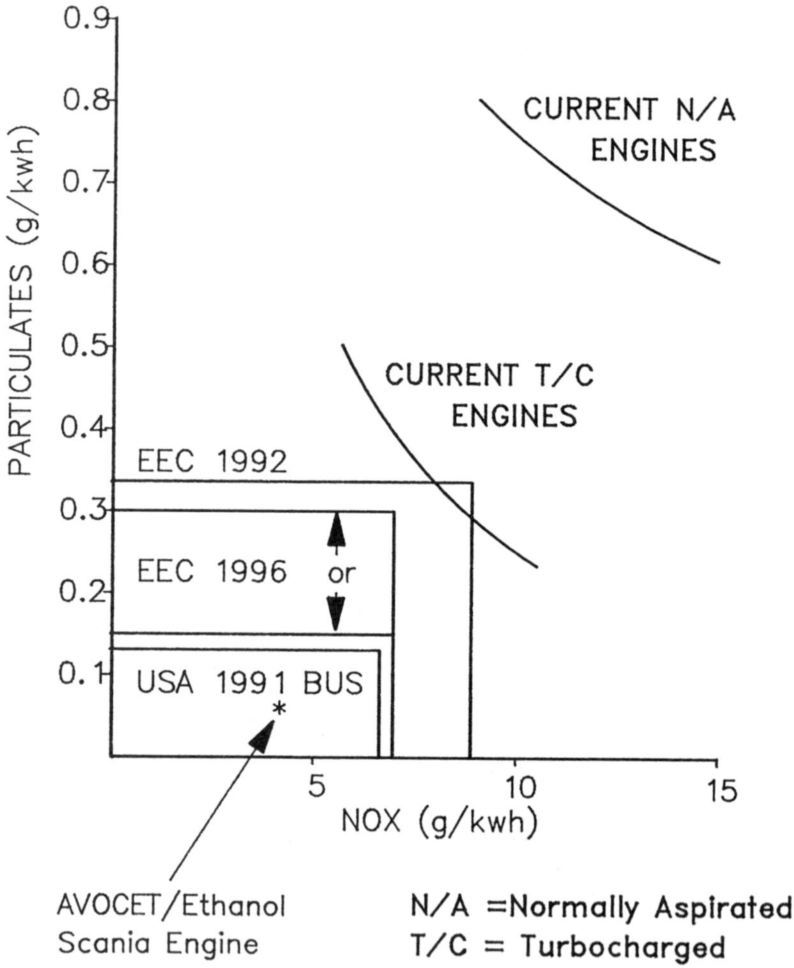


FIGURE 1 **HEAVY DUTY DIESEL ENGINE EMISSIONS**

a) turbocharging and aftercooling,
b) high pressure fuel injection equipment (FIE) with electronic control,
c) re-matched combustion systems,

has produced results which approach the 1991 bus limits.[1] In order to meet these standards however, it is thought necessary to implement further modifications:

a) four valve combustion system,
b) reduced oil consumption,
c) variable geometry turbocharging (VGT),
d) advanced transient control of VGT and FIE,
e) and possibly exhaust after treatment e.g. traps.

Clearly engine costs will be higher and the technology is reaching the limits of its capability, with no potential for further emission improvements. There will be a fuel consumption penalty of 5 to 10% associated with achieving the very low NOX levels. Furthermore the cost of removing sulphur from diesel down to a level of 0.05% has been estimated at 3 to 6 cents per US gallon by CONCAWE,[2] and at 10 to 25 cents per US gallon by others. It is not surprising, then, that there has been an increase in interest in alternative fuels and in technology which could enable the new standards to be achieved by modification or retrofit of existing engines rather than the much more expensive provision of completely new units.

## 4 ALTERNATIVE FUELS

The front runners in the alternative fuels race are ethanol, methanol and compressed natural gas. All these have already demonstrated significant environmental benefits as replacements for diesel fuel as well as for gasoline. The lack of sulphur in these fuels, and the absence of particulate formation, mean that gaseous emissions can be further reduced, in comparison with diesel fuel, by installation of catalytic converters.

Methanol is made from natural gas but is a much more convenient form of transport fuel avoiding the need for cumbersome high pressure storage at the filling station and on the vehicle, and the associated expensive refuelling facilities. Reserves of natural gas, the preferred feedstock for methanol, are plentiful in many parts of the world, offering reduced dependence on the Middle East as the primary source of the world's energy. If currently vented or flared natural gas were used to produce methanol on a large scale, a significant contribution to global warming would be reduced.

The price of bulk chemical grade methanol is currently about 30 to 40 cents per US gallon. A report issued by the US Department of Energy estimates that fuel methanol could be made available on a large scale in the future at less than 30 cents per US gallon.[3] The benefits offered by ICI's Leading Concept Methanol technology could further reduce the costs of production by 6 cents per US gallon. Current quality diesel fuel prices in 1989 varied between 46 and 63 cents per US gallon, which is equivalent to 20 to 28 cents for the quantity of diesel with the same energy content as one gallon of methanol. It is clear that the cost of methanol could at worst be twice the cost of diesel, and at best be no more expensive. However, it should be noted that the cost of fuel represents only 5 to 10% of the cost of running a fleet of buses.

Ethanol is currently expensive to produce from bio-feedstocks, but offers the potential in the longer term of being one of the few renewable fuels which will not contribute to global warming. In the meantime its use as an alternative fuel depends on government subsidies or special local economics.

## 5 IGNITION IMPROVER FOR ALCOHOL FUEL

In order to burn alcohol in a diesel engine it is necessary to modify either the engine or the fuel. The former involves the installation of a spark or glow plug and is the mechanical solution, but it does incur additional costs, and is virtually impossible to retrofit an existing engine. The latter involves the addition of an ignition improver to the fuel and represents the chemical solution. The requirements for a fuel containing such an ignition improver are as follows:

a) low ignition improver concentration,
b) small increase in price compared with neat alcohol,
c) stability, safety and low toxicity,
d) low corrosivity, especially to engine fuel system,
e) no adverse effect on emissions.

Many of the available chemicals, mostly nitrate esters, have been tested as ignition improvers. As long ago as 1981[4] the following results were reported as the minimum ignition improver content in ethanol fuel to obtain the same quality as diesel fuel:

| Ignition improver | Minimum concentration (% v/v) |
|---|---|
| isopropyl nitrate | 25 |
| isoamyl nitrate | 16 |
| hexyl nitrate | 12 |
| 2-ethyl hexyl nitrate | 12 |

| | |
|---|---|
| diethylene glycol dinitrate (DEGDN) | 6 |
| triethylene glycol dinitrate (TEGDN) | 4 |

More recently ICI has introduced AVOCET, a purpose-designed ignition improver. The minimum concentration for AVOCET in the same engine is estimated at 3.5%. Clearly it is at least as effective as the best of the other additives tested, and it is not unstable as in the case of DEGDN or TEGDN. For obvious reasons it is not possible to disclose the exact chemical composition of AVOCET except to say that the active component is also a nitrate ester. However, AVOCET also contains lubricity and anti-corrosion additives to ensure compatibility with the engine fuel supply system.

In the case of the chemical solution the traditional benefits of the diesel engine simplicity, durability and low maintenance costs are retained. The homogeneous combustion produced by pure compression ignition is both environmentally cleaner and more fuel efficient than that produced by electrical ignition. Existing engines can be retrofitted so that the environmental benefits can be achieved without waiting for existing vehicles to be replaced. All that is required is to up-rate the fuel supply system to handle the larger volume of alcohol required, and to replace those components e.g. gaskets, seals and aluminium piping etc. which may be susceptible to attack by alcohol. In the case of a pilot trial in two buses in Auckland (see below) the conversion costs were US $6,000 per bus, but for larger numbers of conversions the cost could be reduced significantly.

## 6 AVOCET IN ACTION

The AVOCET system has proved durable in service with both methanol and ethanol. In New Zealand two MAN-engined buses fuelled with AVOCET/methanol covered a total of 220,000 km on regular service in the city of Auckland between July 1984 and July 1987. Fuel consumption demonstrated an increase in thermal efficiency of up to 15%. Performance and drivability were the equal of the diesel version. There were no mechanical failures attributable to the change of fuel. Engine durability, assessed on the basis of engine oil analysis, was judged to be equivalent to the diesel engine.[5]

In France four Renault-engined buses were converted to AVOCET/ethanol by the Orleans company Moteurgaro. These buses have completed a total of over half a million miles on regular service in the city of Tours since December 1987. The availability of the buses has been equal to their diesel counterparts. Engine noise and vibration has been found to be reduced. One of the engines, which was stripped down and inspected after 81,000 km (about 5,000 h), showed no signs of wear. The emission results from the Tours buses, achieved without the use of a catalytic converter, are shown in figures 2 to 5. Particulate emissions in g $kwh^{-1}$ are estimated from the opacity measurements which showed a reduction from 0.15 $m^{-1}$ to less than 0.02 $m^{-1}$.

In the USA, ICI is jointly sponsoring an AVOCET/methanol trial (project MASSCAR) with the South Coast Air Quality Management District, the agency responsible for air quality in the Los Angeles basin. The object of the trial is to demonstrate the retrofit capability of the system on up to twelve two-stroke Detroit diesel buses. The results from the second generation of retrofitted engine with a catalytic converter are shown in figures 2 to 5.

On the18th April 1990 a fleet of fifteen AVOCET/ethanol-powered Scania buses was launched in the city of Stockholm. A further fifteen buses will be added in September 1990, as part of a £9m programme funded by Stockholm Transit, the Swedish Goverment, Saab-Scania and Swedish Ethanol Chemistry (SEKAB), who provide the fuel. The object of the programme is to assess the environmental benefits of alternative fuels. These buses, equiped with high (24:1) compression engines and catalytic converters have already achieved emission results well within the EPA limits running with concentrations of AVOCET as low as 2% v/v, figures 1 to 5.

The Daimler-Benz results shown in figures 2 to 5 are estimated on the basis of bench tests carried out using an ignition improver similar to AVOCET.[6] Other AVOCET/alcohol trials are taking place in North America as part of the multi-million dollar Methanol in Large Engines (MILE) programme which includes two Cummins-engined garbage trucks in Vancouver, and a shuttle bus at the J.F. Kennedy Airport, New York. Further afield, successful trials have been completed at the University of Santa Maria in Santiago, Chile, and in Taiwan.

## 7 SUMMARY OF EMISSION RESULTS

Particulate emissions are within the EPA limits in all cases, figure 2. There is a significant reduction in NOX emissions in every case, figure 3. Low NOX emissions are achieved because of the lower combustion temperature, and because the nitrogen contained in AVOCET is emitted as elemental nitrogen rather than as NOX. The EPA NOX limit has not been met in every case, but it is possible to reduce further NOX emissions without the increase in particulates associated with diesel powered engines. Hydrocarbon emissions are on the EPA limit with no catalyst, but can be brought well within the limit by the installation of a catalytic converter, figure 4. Carbon monoxide emissions from both diesel and alcohol powered engines are well inside the EPA limits, and can

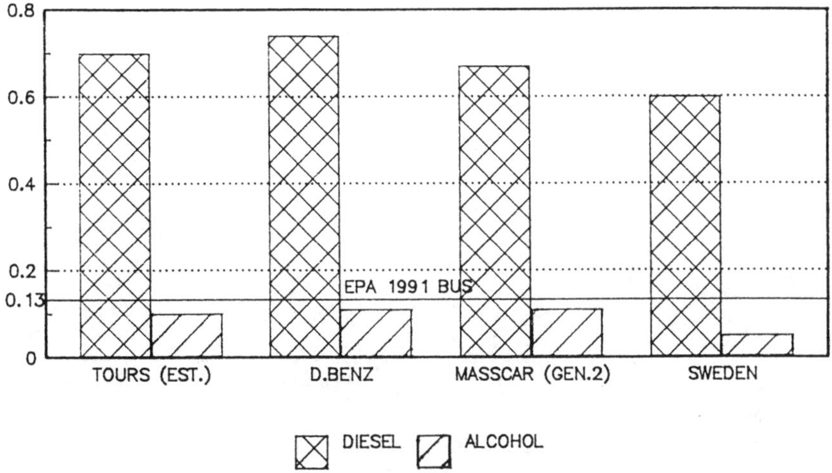


FIGURE 2 PARTICULATES (g/kwh)

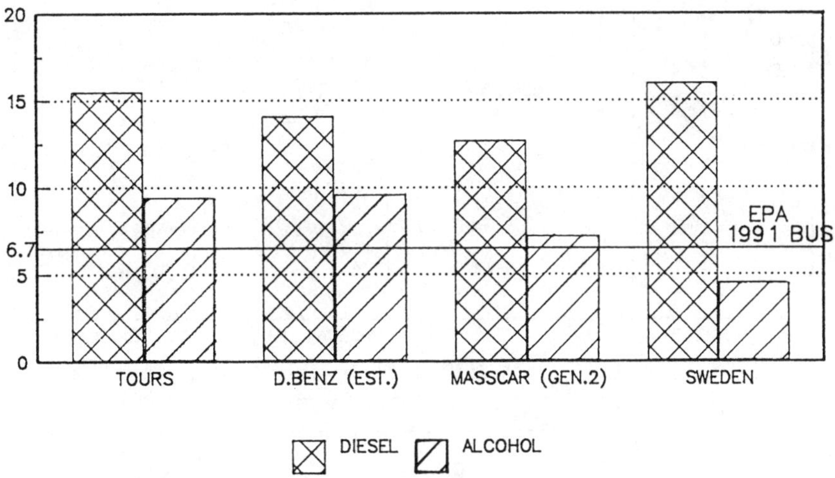


FIGURE 3 NOX (g/kwh)

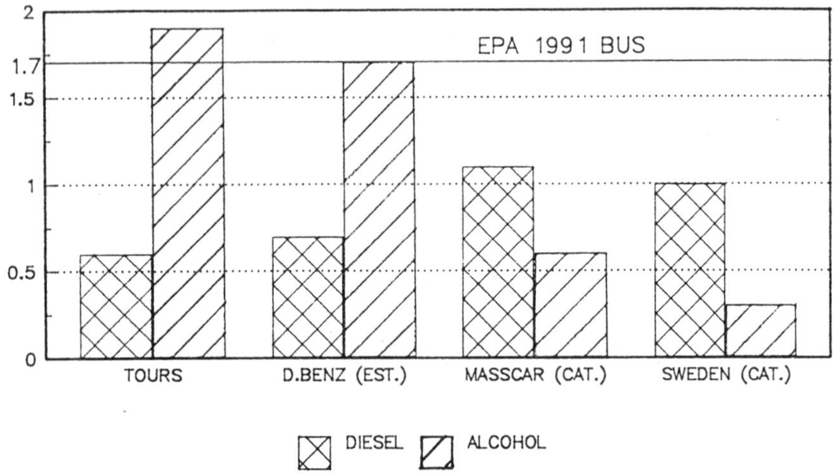


FIGURE 4 HYDROCARBONS (g/kwh)

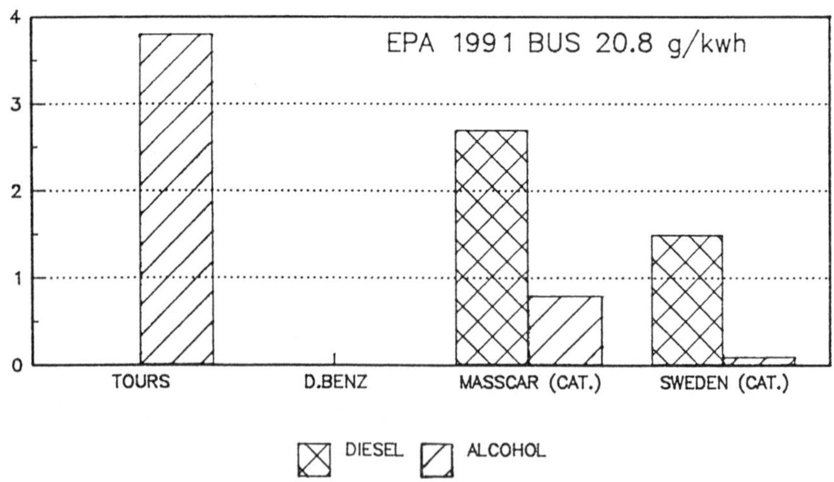


FIGURE 5 CO (g/kwh)

be virtually eliminated with a catalytic converter, figure 5.

Preliminary results on aldehyde emissions from the methanol engines indicate that they are at a similar level to emissions from diesel engines. The potential environmental benefits from AVOCET/alcohol are therefore clear.

## 8 CONCLUSION

AVOCET/alcohol, as a replacement for diesel fuel in CI engines, provides a way of meeting the EPA emission limits for four-stroke and two-stroke engines, both new and as a retrofit. The system is mechanically reliable and durable with driving performance as good as diesel.

The economics depend on the relative costs of alcohol, especially methanol and diesel. In the short term alcohol fuels are likely to cost 50 to 100% more than diesel, though this penalty would be reduced by the use of advanced technology in the new plants needed to produce the required methanol for large scale use. Since fuel represents only 5 to 10% of the cost of running a fleet of buses, even the short term case would add only 2.5 to 10% to the cost of the fleet operation.

The combination of AVOCET and alcohol therefore represents a practical way of reducing atmospheric pollution from diesel engines to acceptable levels. Even with the short term higher cost of alcohol fuel, the off-setting low retrofit cost of the AVOCET/alcohol system means that specific problems of urban fleet emissions could be tackled.

A widespread transition to alternative fuels will need to be co-ordinated over a longer period, with the necessary investment in capacity for methanol and/or ethanol. The timescale needed for this would encompass the associated activities of modification of fuel distribution facilities, provision of the larger quantities of AVOCET needed and programmes for conversion of the large number of engines involved etc.

ICI is continuing to work with a variety of organisations, environmental agencies, urban and mass transport authorities, and engine manufacturers to demonstrate and improve the technology, and to find ways to facilitate its implimentation to make what is believed could be a major contribution to the improvement of our environment.

## REFERENCES

1. J.R. Needham et al, 'Developing the Truck Engine for Ultra Low Emissions', presented at Institute of Mechanical Engineer's seminar, 1st February 1990.
2. CONCAWE report no. 10189, 'Costs to Reduce the Sulphur Content of Diesel Fuel'.
3. United States Dept of Energy, DOE/PE-0093, 'Assessment of Costs and Benefits of Flexible and Alternative Fuel use in the US Transportation Sector', technical report number 3, Methanol Production and Transportation costs, November 1989.
4. H.O. Hardenburg and A.J. Schafer, SAE 811211, 'The Use of Ethanol as a Fuel for Com pression Ignition Engines'.
5. Auckland Regional Authority, Liquid Fuels Trust Board, project no. 611/04/03, 'Ignition-improved Methanol Diesel Trial', July 1984 - July 1987.
6. H.O. Hardenburg, SAE 872093, 'Comparative Study of Heavy Duty Engine Operation with Diesel Fuel and Ignition-improved Methanol'.

# AQUEOUS AND NON-AQUEOUS ENGINE COOLANTS BASED ON PROPYLENE GLYCOL

R.G. Dingley

ARCO Chemical Europe, Inc
Maidenhead
Berkshire, SL6 1YP
United Kingdom

## 1 INTRODUCTION

Engine coolants form a vital function in most internal combustion engines. The reason for this is that only about a third of the energy liberated by burning fuel is used to generate motive power, another is dissipated in the exhaust and the rest heats the metal of the engine. This paper discusses the use of monopropylene glycol (MPG) as the antifreeze component of a modern coolant and briefly summarises the HYBRID™ system in which the entire coolant is MPG.

## 2 CHARACTERISTICS OF ENGINE COOLANTS

If the heat generated in the metal of the engine were not removed, the engine would finally seize as the lubrication system failed. The main function of the cooling system, therefore, is to transfer heat from the metal of the engine to the radiator so that it can be safely dissipated to the atmosphere. To carry out this function the fluid should have a high specific heat, be liquid over the entire operating range of the engine, have a high boiling point and low freezing point, it should not corrode the metals of the engine, it must be chemically stable, it should not be flammable or toxic and of course it should be readily available and inexpensive.

No single material can satisfy all these demands but fortunately water comes closest with high cooling power and good heat transfer characteristics.

Unfortunately, it also has several deficiences. In particular it freezes at too high a temperature and boils at too low a temperature and in the presence of air, it is highly corrosive to metals.

To overcome the high freezing point, most people know that a glycol is dissolved in the coolant water but in a commercial antifreeze concentrate, there can be up to ten other ingredients which act to prevent corrosion of the metals of the engine and reduce the degradation of the glycol. With the increased performance of modern vehicle engines, manufacturers are developing more stringent requirements for the coolants and furthermore there is increased awareness of the need to use safe, non-toxic products whenever possible in any industrial or consumer environment. For these reasons MPG is being increasingly considered, not only as a viable antifreeze product, but as the preferred product where high performance and safety to the public are concerned.

## 3 PROPYLENE GLYCOL AS AN ANTIFREEZE

Since the main function of the glycol is to reduce the temperature at which the fluid ceases to function, it is instructive to compare the freeze point of ethylene glycol (EG) and MPG solutions in water at various concentrations. Figure 1 shows the freeze point measured using ASTM D-1177 for volume concentrations of glycols from 0 to 60%. The difference in freeze point measured by this method is about 2.5 C over the useful range of concentrations. The range bars on figure 1 show the range of values obtained when the same MPG solutions were tested by two different operators using three different examples of a density freeze point tester calibrated for MPG. It can be seen that the range is up to ±3 C indicating that the accuracy of measurement is within the difference of freeze temperature of the two glycol solutions. Furthermore, there are few places in Europe where protection to greater than -30 C is required so that normal glycol concentrations of 50% would provide adequate freeze protection for most circumstances.

For these reasons there is generally no need for the concentration of MPG based antifreeze solutions to be any greater than EG systems unless exceptional conditions are expected when higher concentrations may be used without difficulty.

## 4 CORROSION BEHAVIOUR

Corrosion protection of metals in an engine is very largely determined by the inhibitors used in the coolant mixture. However, the type of glycol does have

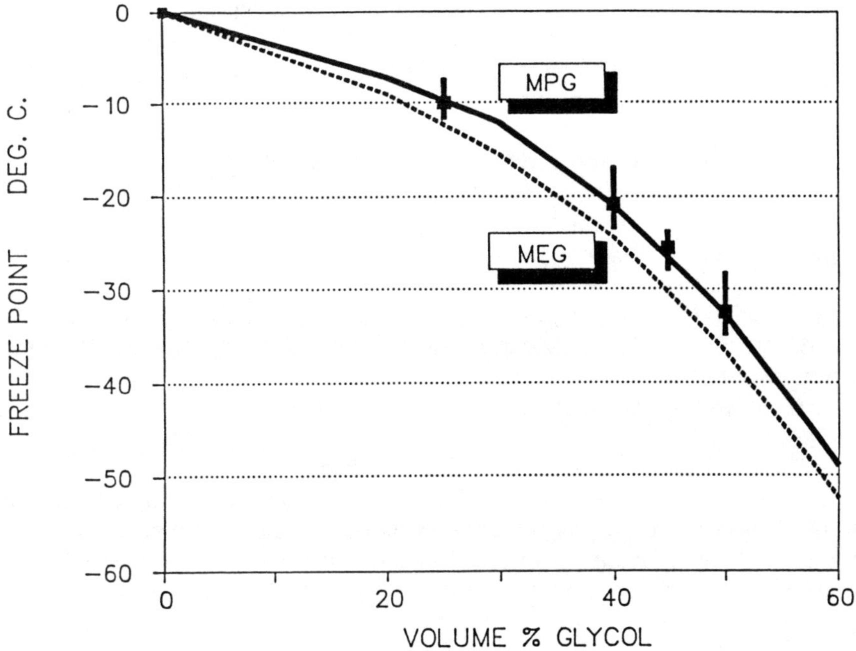


FIG. 1 - Showing ASTM D-1177 Freezepoint as a function of Glycol Concentration

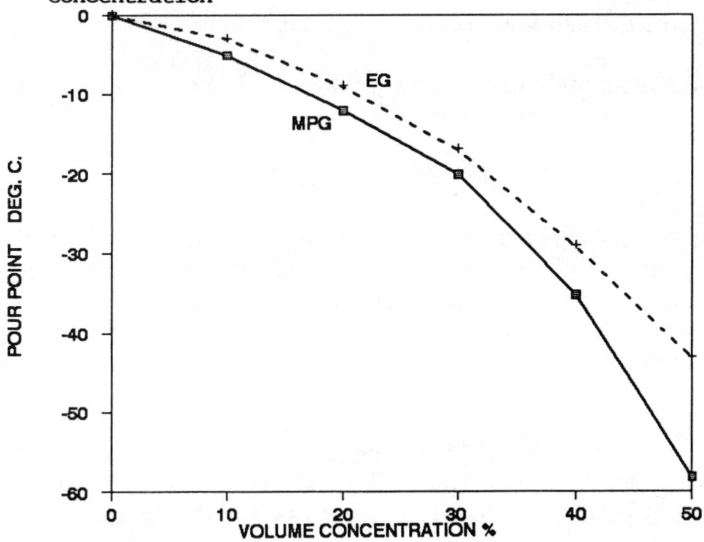


FIG. 2 POUR POINT OF MPG AND EG SOLUTIONS

a significant effect on the type of corrosion caused by cavitation of the coolant and which results in severe pitting of metal surfaces. This problem is particularly severe on the wet cylinder liners of heavy duty diesel engines and on the impellers of some water pumps.

In both laboratory and engine tests comparing the two types of glycol, MPG has been shown to give a significant improvement in protecton against cavitation corrosion of cast iron and aluminium. Hercamp, Hudgens and Coughenour[1] carried out laboratory tests using a cast iron probe vibrating at 20 kHz and 1.25 x $10^{-3}$ amplitude to generate cavitation conditions in 15% commercial coolant solutions. They compared the weight loss after 6 days at 71 C for EG and MPG based systems both with the same inhibitor package. Their results indicate a clear advantage for the MPG based system which resulted in a weight loss of only 6.3 mg compared with 23 mg for the EG coolant.

The results from these tests were confirmed in running a modern, high powered diesel engine on a dynamometer where the MPG coolant was shown to eliminate completely any pitting of the cylinder liners. They concluded that the use of MPG provides extra cavitation corrosion protection equivalent to at least 20 to 40% more than the inhibitors in a conventional formulation. Other points to emerge from these tests were:

a) engine temperature, fuel consumption and other performance parameters were similar for both systems,
b) MPG is more resistant to oxidative degradation than EG and the coolant may therefore last longer,
c) there is a reduction in the corrosion of aluminium when tested under heat rejection conditions (ASTM 4340).

## 5 SAFETY AND ENVIRONMENTAL ISSUES

There are increasing demands from consumers for safer and environmentally acceptable products to replace those which are either dangerous or pollute the atmosphere. This is particularly strong for chemicals which have a generally poor image in the consumers' mind. If safer alternatives exist, therefore, it must be right to use them. It is in this respect that MPG shows an important benefit over EG.

EG is classified by the FDA and the EEC authorities as harmful and as a result is subject to OSHA and COSHH regulations. In the EEC any antifreeze made with EG has to carry the St Andrew's cross indicating it to be a dangerous product. This is no scare story as there are recorded instances of human deaths from drinking antifreeze. In the USA over 50% of all deaths from poisoning of cats and dogs are attributed to EG. The American Occupational Safety and Health Administration requires that exposure to EG does not exceed 50 ppm. This value was "Based on evidence of an occupational risk of severe throat and respiratory irritation". In the EEC the limit is 60 mg $m^{-3}$ for vapour and 10 mg $m^{-3}$ for a mist of suspended particles with a maximum of 125 mg $m^{-3}$ over a ten minute period. In contrast, MPG is classified as "Generally Regarded as Safe" by the FDA and is widely used in petfoods, cosmetics and detergents as well as being an antidote for EG poisoning.

In recognition of this difference in toxicity, the Swiss Poison law of 1972 banned the sale of EG antifreeze in supermarkets, shops and generally across the counter to the public. Austria introduced a smilar law in 1989 after problems with poisoning from EG in wine. For this reason MPG antifreeze concentrates are widely used in these countries and have given no problems.

MPG is readily biodegradable and is classified as relatively harmless to freshwater and marine algae, invertebrates and fish according to a toxicity rating of the US Fish and Wildlife Service. It has no known detrimental affect on the environment.

ARCO Chemical Company is a leading world wide producer and supplier of propylene glycols with factories in Europe, Asia and the USA and sales organisations in all major markets. MPG is a world scale commodity chemical with current capacity approaching 1 million tonnes. Availability is therefore no problem.

MPG based coolants are available now and have gained wide acceptance in Italy as well as Switzerland and Austria. The growing awareness of the advantages it affords, from both a technical and safety point of view, will ensure it is used more extensively in future. After all:- *if there is a safer alternative available, why not use it?*

## 6 NON-AQUEOUS OR HYBRID™ COOLANT SYSTEMS

No discussion of MPG based engine coolants can be complete without some mention of a very interesting development where all water is eliminated from the system. Such a system has been developed over the last ten years by Jack Evans of MECA Developments Inc. and Tom Light of National Technologies Inc. in the USA. This subject merits a whole paper to itself as indeed the above gentlemen have presented to many distinguished audiences. However, this paper summarises the most important aspects of the system.To understand the reasoning behind the idea it is necessary to review the mechanism of heat transfer from the engine to the cooling system and subsequent dispersion.

There are two principal mechanisms at work; convection and nucleate boiling. Convection is the mechanism operating in the radiator and to some extent in cooling the engine block in the cooler areas. The efficiency obviously depends upon the specific heat, conductivity and viscosity of the medium; water is very good in this respect.

When the temperature of the metal approaches the boiling point of the cooling fluid localised vaporisation occurs at the interface between the metal and the fluid followed by immediate condensation of the vapour in the bulk fluid. This vaporisation and condensation (referred to as nucleate boiling) transfers very high quantities of heat to the fluid since, of course, the heat of vaporisation of a liquid is much higher than the specific heat.

Problems arise with this method of heat transfer when the engine operating temperature is close to the boiling point of the coolant for a given pressure. Under these conditions the vapour cannot condense efficiently and vapour accumulates at critical areas of the interface causing a vapour blanket and catastrophic reduction in heat transfer. Other problems result from flash vaporisation in the water pump causing loss of fluid circulation. After boil is also likely after the engine is shut down and residual heat dissipates into stationary coolant. For these reasons water based systems have to run at relatively low temperatures or under pressure to increase the boiling point.

Jack Evans conceived the idea of replacing water with a fluid having a much higher boiling point than water but which also had good heat transfer characteristics and did not freeze under normally encountered conditions. After considering a number of alternatives he concluded that MPG showed the best overall properties for his purpose. Some properties of MPG and a 50/50 mixture of EG and water are shown below:

| Property | MPG | 50/50 EG/water |
|---|---|---|
| boiling point | 187 | 107 |
| freezing point | -57 | -37 |
| heat of vaporisation (kcals mole$^{-1}$) | 12.5 | 9.8 |

The much higher boiling point and lower freezing or pour point of MPG are evident advantages. The high boiling point achieves the objective of a large difference in temperature between the coolant and vapour temperature resulting in rapid condensation of any vapour formed. This ensures liquid is always in contact with the metal surfaces and reduces the possibility of hot spots being formed.

Of more significance is the high heat of vaporisation per mole of MPG compared with the water mix. For a given quantity of heat absorbed by vaporisation, this means that there is less vapour formed which in turn means there is again less chance of vapour blanketing or cavitation. Other advantages of the HYBRID™ system include:

a) no cavitation in the coolant pump resulting in longer life and reduced maintenance,
b) the flow of the coolant can be reversed giving more efficient cooling at the cylinder head where it is needed most and increasing the temperature of the cylinders so promoting more efficient fuel combustion,
c) elimination of cavitation corrosion of cylinder liners,
d) the coolant is at atmospheric pressure reducing the stress on hoses and seals. This is a major contributor to reduced maintenance of the engine,
e) antifreeze and supplementary coolant additives do not have to be used,
f) there are no dissolved minerals such as calcium salts which form scale on water systems and reduced thermal conductivity,
g) coolant temperatures can be increased and the radiator size reduced resulting from the more efficient heat transfer at the higher temperatures.

HYBRID™ coolant system has been extensively tested in laboratory and field trials. The main advantages that have emerged from these trials are considerably greater reliability of the engine as a result of the reduced coolant pressure and lack of corrosive material in the system. In racing engines running at maximum output for several hours there is a greatly reduced incidence of cylinder head cracking and corrosion from hot spots.

The patented system is still being developed by MECA Industries and National Technologies who are now marketing an inexpensive kit for trucks and high performance cars so that they can achieve maximum benefit from the system. Further information is available from Mr J. Evans, (MECA Developments Inc., Route 41 RR140, Sharon CT, USA 06069) or Mr T. Light (National Technologies, P.O. Box 1479, Sharon CT, USA 06069).

## 7 CONCLUSIONS

MPG has been shown to be an acceptable alternative to EG for use as a freeze point depressant in automotive coolants. It has distinct safety advantages over EG and technical advantages especially when used in heavy duty engines.

The complete elimination of water in engine coolants by using 100% MPG has already proved beneficial in high powered racing engines. With the drive towards longer lasting sealed coolant systems, this may well be the preferred method for all cars of the future.

## 8 ACKNOWLEDGEMENTS

I should like to thank ARCO Chemical Company Inc. for permission to publish this paper and my colleagues, Glen Coughenour, Bill Weeks and Jim Greaney for help in its preparation.

## REFERENCES

1. Hercamp, Hudgens and Coughenour, 'Aqueous Propylene Glycol Coolant for Heavy Duty Engines', Proc. Int. Congress, Detroit, 1990, SAE Techical Paper SP 811-900434.

# SYNTHETIC RESINS FOR GLASS-FIBRE REINFORCED PLASTICS

Dr P.R. Darby

Cray Valley-Total
Stallingborough
South Humberside DN37 8DR
United Kingdom

## 1 INTRODUCTION

The use of plastics in vehicle construction has increased progressively over the last fifteen years, now forming some 8 to 12% of the weight of an average passenger car. The level of adoption of plastics varies according to manufacturer and model and is illustrated in table 1:[1]

Table 1 Plastic content of some popular motor cars

| model | kg plastic | %vehicle weight |
|---|---|---|
| Peugeot 205 | 72 | 8 |
| Citroen BX | 102 | 11 |
| Ford Sierra | 120 | 13 |
| Mercedes W124 | 100 | 8 |
| BMW 3 series | 99 | 9 |
| VW Golf | 59 | 7 |

A breakdown of the nature of the plastics used and indicative trends can be illustrated by the table 2:[2] It should be stressed that although the main part of these plastics can be attributed to the interior trim not structural, re-inforced applications, they do indicate the increasing acceptance of plastics and composites by the somewhat conservative automotive industry.

Table 2 Changing use of particular plastics in modern cars

| material | 1980 | 1985 | 1990* |
|---|---|---|---|
| PVC | 12.5 | 11.0 | 8.0 |
| PU | 11.2 | 15.0 | 20.0 |
| PP | 8.0 | 11.0 | 15.4 |
| ABS | 6.4 | 6.4 | 6.5 |
| nylon | 2.75 | 3.5 | 5.8 |
| polyester compounds | 1.50 | 5.5 | 11.0 |
| phenolic compounds | 7.0 | 5.0 | 3.5 |
| polyethylenes | 2.5 | 4.0 | 11.6 |
| polycarbonate | 0.3 | 3.0 | 5.3 |
| mod: PPO | 0.3 | 3.0 | 4.1 |
| acetal | 0.8 | 2.0 | 5.85 |
| RIM PU | - | 1.2 | 2.4 |
| PBT | - | 0.5 | 1.0 |
| cellulosic | 0.2 | 0.15 | 0 |
| acrylic | 1.2 | 1.0 | 0.7 |
| others | 0.02 | 2.5 | 3.6 |
| TOTAL | 54.8 | 74.25 | 104.35 |

* projected

## 2 BACKGROUND

This paper concerns the use of polymer composites in structural automotive applications, and more specifically, those composites derived from glass-fibre reinforced polyester resins. 'Structural' applications can be classified as follows:

| | |
|---|---|
| semi-structural | doors, bonnet, floor panels |
| structural | chassis, frame |
| zero-failure | springs, steering rack, engine components. |

The suitability of polyester composites as regards each of these functions will be discussed and illustrated later.

The advantages of using composites in structural automotive applications are well documented and can be summarised thus:

Weight Reduction. Readily transferable into improved fuel economy and performance. As a rule of thumb, replacement of a steel component with composite results in a 50% weight saving.

Three Dimensional Mouldability. The ability to provide complex shapes in a single moulding operation reduces labour and robotic requirements, and increases stiffness/performance.

Corrosion Resistance. Quite simply plastics and composites do not rust, extending vehicle life, reducing maintenance and improving safety.

Energy Absorption. Low speed collisions result in less damage than with steel.

Capacity to Damp Vibration and Noise. Improves comfort and ride.

Styling Freedom. Particularly important with low/medium production volumes, that is specialist and niche vehicles.

## 3 SPECIFICATION OF POLYESTER COMPOSITE

Unsaturated polyester composites represent by far the most abundant polymer composite in structural automotive applications. Table 2 above indicates that in 1985 an average passenger car contained a substantial 5.5 kg polyester composite. It should be stressed at this point that polyester has also found a significant use in commercial vehicles. The general reasons for specifying polyesters can be summarised as follows:

Ease of Handling. Resin is easy to spray and inject, tolerant to handling variation and relatively non-toxic.

Versatility. In terms of polymer formulation, additives and fabrication techniques.

Mechanical Properties. Excellent portfolio of properties particularly in comparison with other plastics.

Rapid Cure. Moulding times of one minute attainable with compression moulding.

Dimensional Stability. Provision of stiffness and rigidity.

Electrical Properties. High resistivity and dielectric strength.

Colour. Light colour and facile pigmentation.

Cost. At about £1 000 per tonne for general purpose polyesters, this compares favourably with other thermosets and some thermoplastics.

## 4 MANUFACTURING TECHNIQUES

The versatility of polyester resins lend themselves to an array of fabrication/conversion techniques running from the labour intensive, low tooling cost hand lay-up to the involved, high tooling cost compression moulding capable of very high production rates. The requirement of any conversion process for automotive parts is taxing and generally requires: high production rates, good part strength, complex parts, low operator skill, high part reproducibility, and ideally low tooling cost. The processes that are commensurate with this specification will now be outlined in further detail.

### Compression Moulding

The compression moulding of sheet moulding compound (SMC) currently represents the nearest match to the automotive requirements. Of particular note are the high production rates and part reproducibility attainable. The technique is established technology with an estimated 30 to 40 000 tonnes per annum automotive use in Western Europe. Well known examples of application include the ERF cab unit and Citroen BX bonnet.

The nature of the chopped glass used in moulding compounds restrains part mechanical strength and hence the process is limited to semi-structural applications. The high pressure of the moulding process limits part complexity and integration. Process improvements sought include better mechanical properties, reduced pin-holing, improved gloss/class A mirror finish, and faster cycle times.

### Resin Transfer Moulding (RMT)

RMT again represents established technology but differs from compression moulding in terms of significantly lower tooling costs. RTM is currently used for specialist car manufacture, for example Lotus, but is characterised by long cycle times of the order of several minutes. RTM is characterised by excellent structural properties and the high part complexity attainable.

There is an opinion that with the necessary development, RTM could provide the method for the volume production of structural automotive components. Progress is required in terms of glass-fibre technology, faster processing resins and computer control of mould filling. Cray Valley are active in all areas of development.

Glass-mat Re-inforced Thermoplastic (GMT)

Structural polymer composites are not confined to thermoset matrices. GMT, primarily polypropylene based, is increasing in use for semi-structural non-visible applications, for example floor panels. Application is currently limited by high cost, limited part complexity and textured finish. However, GMT can be processed rapidly by flow moulding or stamping, and so has resulted in a warm reception from the automotive industry.

## 5 COMMERCIAL APPLICATIONS

Commercial Vehicles

Commercial vehicle manufacturers have been stalwarts in specifying polyester composites from the early 1970s. The ERF tractor cab can be regarded as the show-piece being manufactured almost entirely from polyester SMC to the tune of about 300 kg composite. The Foden cab is constructed similarly, but by a hand lay-up process. The substantial use of polyester composites in commercial vehicles is attributed to the lower volume production runs and hence the need to moderate tooling costs, and the less critical need to provide a class A mirror finish. More generally, polyester composites find extensive use in most tractor cabs and commercial vehicles. Applications include aerofoils, bumpers, cab sleepers, doors, fascias, grilles, noise shields, rooves, step panels, wings and wheel arches.

Cars

The automotive market for polymer composites is both fragmented and volatile. The different concepts of application will be illustrated by specific examples.

Specialist Sports Cars. Typified by Lotus and TVR in the United Kingdom. Exterior panels are manufactured by hand lay-up and RTM processes with moulding times of the order of several minutes. The very low runs and required styling freedom results in there being little alternative to polymer composites in these specialist cars. In recent years a number of the major car manufacturers have launched specialist sports cars similarly manufactured substantially from polymer composite. Examples include the Alfa Romeo SZ, Renault Alpine and

BMW M3.

Renault Espace. The Espace 'people carrier' represents the new breed of niche vehicle. All exterior panels are fabricated from polyester composite by RTM and are bolted onto a steel frame. Polyester/RTM was specified to meet the initial low production volumes and simply emulates steel panels. The success of the Espace has resulted in the requirement for higher production volumes which will be met by the commissioning of SMC tooling in the near future.

Citroen BX. The SMC bonnet represents the largest single application of any composite in a mass produced car. Processing entails a three minute moulding cycle, use of a low profile resin to minimise pin-holing, an in-mould-coating, hand-finishing and painting. The C-pillar, grille and rear wheel covers are similarly fabricated in SMC. The tailgate is manufactured by the injection of bulk moulding compound (BMC).

General Motors APV. The All-Purpose-Vehicle (APV) was launched in early 1990 and represents the first mass produced car to encompass all composite exterior panels to the tune of 150 kg SMC. Projected 1990 sales were 250 000. Composite was specified for this integral design to attain maximum part integration and weight saving. The APV does not represent simple substitution of steel panels for composite in a bolt-on manner, but a vehicle that has been designed according to composite engineering principles.

Components

SMC Bumpers. Renault introduced the SMC bumper in the early 1970s and up until the mid 1980s SMC bumpers represented *the* major use of polyester composites in cars. Unfortunately, a move to polypropylene and RIM polyurethane has greatly diminished this application for polyesters. SMC bumpers are still specified for some Renault and Peugeot models, but the use of SMC for structural bumper beams is still widespread.

Headlamp Reflectors. The injection moulding of BMC for these components represents a major application for polyesters. The requirement for critical design and heat resistance is readily met by polyester composites.

Under Bonnet Applications. The much talked about under bonnet applications for polymer composites are now starting to materialise. For example, a number of manufacturers now use polymer composite rocker covers for example Fiat Uno turbo. Advantages include weight saving, noise reduction and the integral co-moulding of the component gasket in a single operation. A number of other composite applications will come to fruition in the next twelve months including inlet manifolds and oil sumps.

## 6 THE FUTURE

The value of the strategic automotive market (including commercial vehicles) in the United Kingdom for polyester composite is currently estimated at £10 million per annum.[3] Although the existing applications should not be dismissed, the rather disjointed series of commercial applications outlined above, serve to illustrate that the widespread use of structural polymer composite in mass produced cars has not yet occurred. The potential large scale adoption of composites can be categorised into three areas, each with particular factors to be addressed.

### Exterior Panels in Specialised/Niche Vehicles

Low/medium production volumes will result in the continued specification of polyester composite in these applications. There is a line of thought in the European automotive industry, that niche vehicles are an effective means of competing with mass produced cars from Japan and the Far East.

### Exterior Panels in Mass Produced Cars

The penetration of polyester composite in this area is very much linked to fuel costs. Escalating fuel costs would see an associated increase in the use of composite with consequent weight saving and fuel economy. In the absence of the above, composite components need to become more cost competitive with steel in terms of material cost and production rates. The key to this challenge being part integration.

The technology most likely to have an impact in this area is the compression moulding of SMC. The required process improvements have been outlined previously, and these will be required to maintain the current advantages enjoyed over thermoplastics. No paper on automotive composites would be complete without comment on recycling. Technical and logistic progress in recycling are essential to guarantee the long term specification of polyester composite.

### Full Structural Components in Niche and Mass Produced Cars

This application represents the logical application for structural composites.

It also represents the most telling challenge with potentially the highest rewards. The requirement is most likely to be met by RTM if moulding times can be reduced. Process improvement is required in terms of glass-fibre technology, faster processing resins and computer control of mould filling. Cray Valley are actively pursuing development in all three areas and with particular emphasis on faster processing resins. Improvement is also required in composite mechanical properties.

Although low energy impact to composite is less than that with steel, satisfactory performance in higher speed collisions has not been attained with standard composite materials. The specification of advanced composites in Formula One racing cars and such like does indicate what can be achieved. In order to succeed in mass production a collaborative effort of material suppliers, composite producers and automotive engineers is required to design-in the advantages conveyed by composite materials.

## REFERENCES

1. National Economic Development Office, 'Plastics in the Automotive Components Market', 1987.
2. N.A. Waterman, *Int. J. of Vehicle Design*, 1985, **SP6**, 1.
3. Cray Valley figures.

# FOAM SANDWICH PROCESSING FOR AUTOMOTIVE APPLICATIONS

P. Atkinson
ICI Advanced Materials
Wilton
Middlesbrough
Cleveland, TS6 8JE
United Kingdom

## 1 INTRODUCTION

Advanced materials and composites are increasingly being applied to automotive applications. The objective is to tailor materials and processes to compete both technically and cost effectively with more conventional metal fabrications. The foam sandwich injection moulding process described here has much to offer. Light but stiff structures with class A paintable skins can be fabricated from cost effective thermoplastics such as polypropylene. Choosing a higher performance polymer (such as ICI Verton™), long-glass re-inforced thermoplastic) results in stiffer and tougher components for semi-structural applications such as boot floors and bulkheads.

## 2 THE PROCESS

Sandwich moulding, sometimes known as co-injection, is a process which was delevoped by ICI. Sandwich moulding is carried out similarly to that of conventional plastic injection moulding except that the machine is equipped with two injection cylinders which sequentially inject two materials into the mould. The first material which is injected partly fills the cavity will eventually end up as the skin of the foam sandwich component, figure 1a. Immediately following this skin the core barrel begins to inject and because of rheological effects this material (injected last), ends up as the core of the sandwich component, figure 1b & 1c. In foam sandwich moulding the core material contains dissolved gas which, because of the pressure maintained on the plastic melt, remains in solution until injection is complete. The tool then 'breathes' expanding in thickness under mechanical control from typically 8 to 12 mm, figure 1d & 2. Gas comes out of solution nucleating and forming a closed cell foam structure which stabilises as the thermoplastic matrix cools and solidifies. The end result is a foam sandwich structure with a skin of one material type and a foamed core of the second material which can have different mechanical and physical properties.

## 3 FOAMING

To produce foaming it is necessary to dissolve gas in the polymer melt within the extruder barrel. It is possible to inject gas directly into the barrel and dissolve it in the polymer, however, it is more convenient to add a chemical blowing agent which decomposes or reacts during the melting phase releasing gas which then dissolves in the thermoplastic melt.

Much has been written on foaming agents and their relative advantages and disadvantages. An ideal blowing agent gas should be:

a) odourless, non-toxic, non-corrosive, non-flammable and chemically inert,
b) soluble in polymer and diffuse more slowly than air through the polymer matrix to aid processing.

Nitrogen, generated by thermal decomposition of azodicarbonamides, and carbon dioxide liberated during the thermally activated reaction of anhydrous citric acid with carbonate (as in Boeringer Ingelheim Hydrocerol range of blowing agents) are two commonly used blowing gases.

## 4 THE RESULT

Foam sandwich moulding combines the better properties of two materials into a single composite component. The foam core of the sandwich reduces the density of the composite but maintains the thickness (and thus the stiffness) of the structure.

For horizontal automotive body panels such as boot lids, bonnets and possibly rooves, the requirement is for a paintable and yet stiff structure with a low coefficient of thermal expansion. We have thus used an easily paintable mineral filled and rubber modified polypropylene (such as ICI Procom™ D1031) and a glass re-inforced (to limit thermal expansion) polypropylene for the core material. The mechanical properties of the component are as follows:

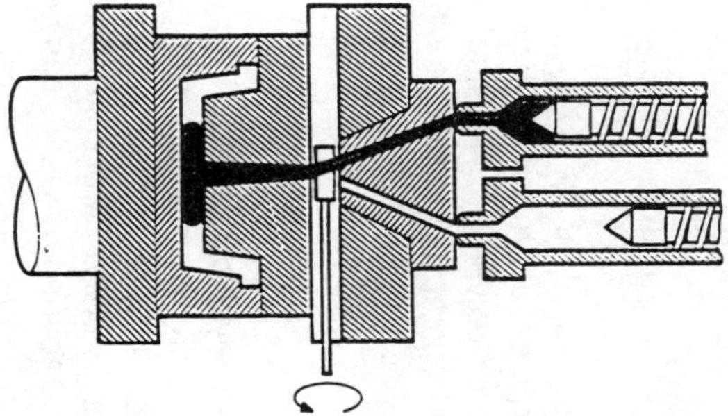

Figure 1a shows the initial stage of mould filling in which skin material alone has been injected, but has not yet formed the skin.

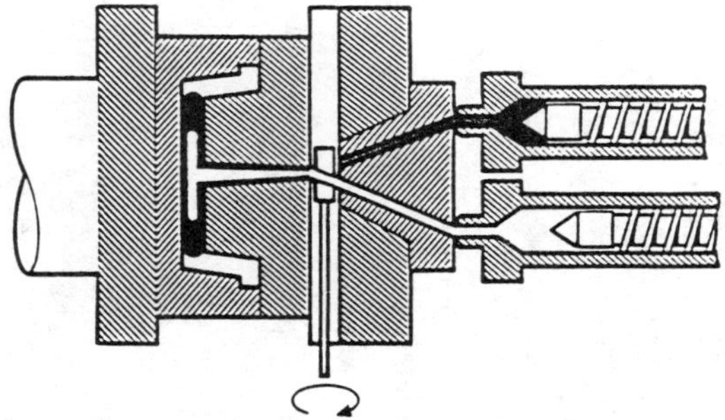

Figure 1b shows injection at an intermediate stage at which core material is flowing laminarly into the interior of the core and has begun to expand the skin as an envelope.

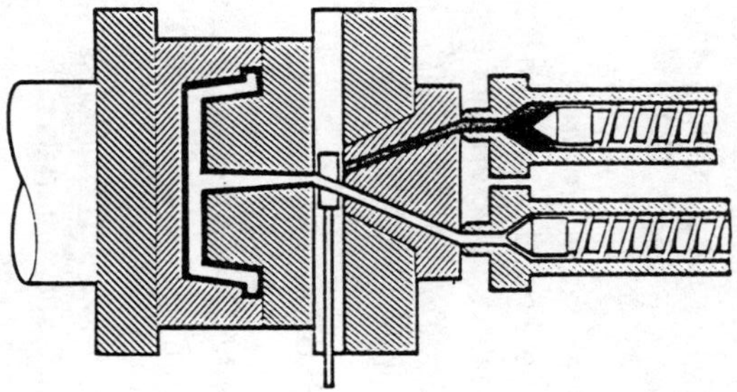

Figure 1c illustrates the completion of injection, about 70 - 90% full in the case of foamcored mouldings, showing that the skin is unbroken and uniformly encapsulating the core.

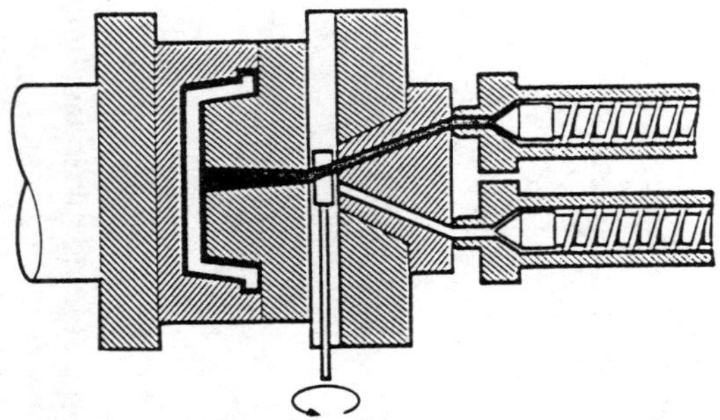

Figure 1d The valve is returned to the start position and cleared of core material in preparation for the next shot. Finally, both polymer streams are isolated from the mould. Cooling then proceeds and the part is ejected.

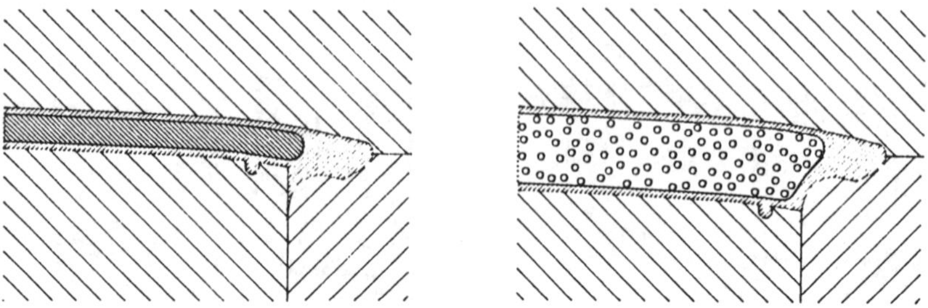

(1) Skin and core materials co-injected into a 6 or 8mm cavity

(2) Cavity expanded to 12mm; blowing agent in core material produces a foam structure

Figure 2 THE PROCESS

| | |
|---|---|
| component starting thickness (before clamp relax) | 8 mm |
| component final thickness | 12 mm |
| composite density | 620 kg $m^{-3}$ |
| effective flexural modulus with flow | 1.2 G Pa |
| effective flexural modulus across flow | 0.9 G Pa |
| thermal expansion coefficient with flow | $3.5 \times 10^{-5}$ $C^{-1}$ |
| (NB because of gating arrangement and fill pattern, 'across', a boot lid is usually 'with' flow) | |
| ductile/brittle transition temperature | -30 C |
| surface quality (class A, DOI) | 9 to 10 |

For applications where the composite structure needs to be stiffer and tougher but where good paint finish is not a vital requirement (for example on a boot floor), we have selected a long glass re-inforced polypropylene skin (ICI Verton™) and a polypropylene core. This results in a composite structure with improved stiffness and toughness but a similar low density.

## 5 CONCLUSIONS

A form of sandwich processing is currently under development at ICI for a number of automotive panel-type applications. The performance of composite structure can be tailored to fit the application by selecting materials and component thickness. The process is environmentally friendly, not only in that it uses thermoplastic materials but also in that recycled thermoplastics can be used as the core material in less demanding applications.

# ENVIRONMENTALLY CONSIDERATE LUBRICANTS FOR THE AUTOMOTIVE AND ENGINEERING INDUSTRIES

S. J. Randles, A. J. Robertson

ICI Chemical Products
Wilton
Middlesbrough, TS6 8JE
United Kingdom

Prof. R. B. Cain

University of Newcastle
Newcastle
United Kingdom

## 1 INTRODUCTION

The issues of ecotoxicity and biodegradability of pollutants are becoming increasingly important features of the growing public awareness of environmental issues. The seventies and eighties have witnessed some important changes in the public perception of environmental problems and they have become much higher on national agendas. The German Federal Republic, Switzerland and the Scandinavian countries in particular have embarked on a series of ever tightening "green legislation" and this trend seems set to spread.

## 2 WHY THE CONCERN

In the European Economic Community (EEC) total sales of lubricants amount to some 4.5 million tonnes per year. Figure 1 shows how this is used: 2.45 million tonnes of lubricants are fully consumed, 700 000 tonnes are reprocessed, 750 000 tonnes are used as fuel and 600 000 tonnes "disappear" uncontrolled into the environment.[1] It is this last figure of 600 000 tonnes that is causing increasing concern because it represents a potential threat to the environment as well as a loss of a natural resource. Every year for instance an estimated 5 000 tonnes of chain-saw lubricant oil seeps into the soil of German forests.[1] Consumption of hydraulic fluids in Germany is 145 000 tonnes per annum (50% in mobile use). Estimations for leakage loss are up to 6 000 tonnes per annum (4%). Even with the installation of better seals and systems this leakage figure is still likely to remain high.[2] These figures are important when one considers that just two pints of motor oil when poured into water can create a slick an acre in area.[3]

## 3 ENVIRONMENTALLY CONSIDERATE LUBRICANTS

For a lubricant to be assessed as environmentally considerate several aspects of its performance should be considered. These are summerised as follows:

a) good biodegradability,
b) low bio-accumulation,
c) low toxicity,
d) low exhaust emissions,
e) good performance.

### 3.1 Biodegradability

Biodegradation may be defined very broadly as the ability of a chemical compound to be broken down by the biological action of living organisms. The concept of biodegradability can extend from a minimal transformation that leaves the molecule largely intact though it may alter some physical properties through partial breakdown of the original molecule to complete mineralisation. Some authorities use the term "environmentally acceptable" biodegradation to indicate a sufficiently extensive molecular cleavage to remove biological, chemical or physical properties associated with the use of the original product.

There are several phenomena associated with biodegradation that can be measured easily using conventional analytical methods,[1] namely:

a) substrate loss,
b) the formation of water or carbon dioxide as end products,
c) oxygen consumption,
d) the release of energy e.g. heat or increase in the number of (biomass) bacteria involved.

Obviously, the most logical way of expressing biodegradation is by substrate loss. The chemist has numerous classic and modern methods at his disposal to demonstrate this from gravimetric establishment of an extracted proportion of oil, through total organic carbon (TOC) and dissolved organic carbon (DOC) analysis (now largely automated), to quantitative infra-red (IR). Chemical oxygen demand (COD) which can be determined by conventional chemical oxidation of all oxidisable water constituents forms part of the German water discharge law. Unfortunately, this method is not precise with water-immiscible substances like lubricating oils.

3.1.1 CEC-L-33-T-82 test. This test was originally developed to characterise the biodegradability of outboard engine oils used on the Bodensee. It evolved as a result of the concern by local authorities and provincial governments, over the accumulation of mineral oil residues in the lake which were thought to be the cause of fish tainting. The test was developed as a Co-ordinating European Council (CEC) tentative test method in 1982 but has only recently become widely used and it is now being applied to other lubricants such as hydraulic fluids.

The CEC test is used to determine the overall level of biodegradability of hydrocarbons, or similar compounds containing chains of methylene ($-CH_2-$) groups. In the test the lubricant is dissolved in a chlorinated solvent and the resulting solution is then introduced into an aqueous media containing sewage inoculum. The mixture is then continually agitated at 25 C in the absence of light for a period of 21 days after which the IR absorbance at 2930 $cm^{-1}$ (which corresponds to the $-CH_2-CH_3$ stretch) is measured and compared to its initial value. To provide cross checks, parallel tests are also run with a poisoned sample and further samples are analysed at 7 and 14 day intervals. Because of the usual water-immiscibility of the oil on test, complete incubation flasks must be sacrificed for each sample; a single incubation flask cannot be progressively sacrificed. Large numbers of replicates are therefore needed. Figure 2 summarises the test method.

Figure 3 shows the biodegradability of various lubricants measured using the CEC test. Mineral oils, polyalkylolefines (PAOs), polyisobutylenes (PIBs), polyethers and alkylated benzenes have poor biodegradability, vegetable oils have good biodegradability whereas esters show a range of biodegradabilities from very high (100%) to very low (5%). Although polyethers show poor biodegradability by the CEC test they do have other advantages, namely that they are water miscible. As they are water miscible the Organisation for Economic Co-operation and Development (OECD) biodegradability tests which are based on oxygen consumption, carbon dioxide evolution or organic carbon removal may be more applicable. By these tests polyethers can be between

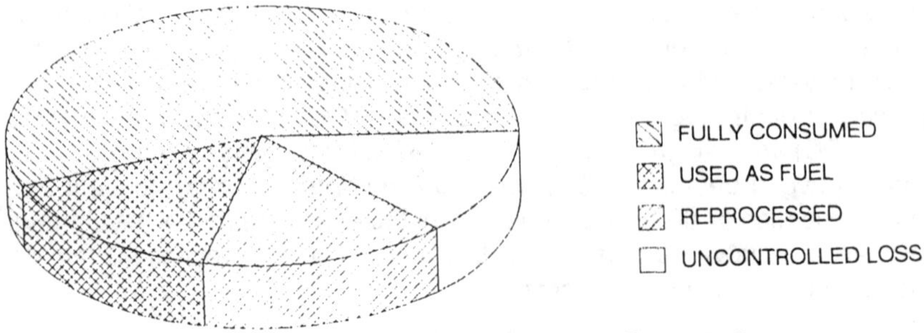


**FIG 1: EEC LUBRICANT SALES**

**( 4.5 MILLION TONS PER YEAR )**

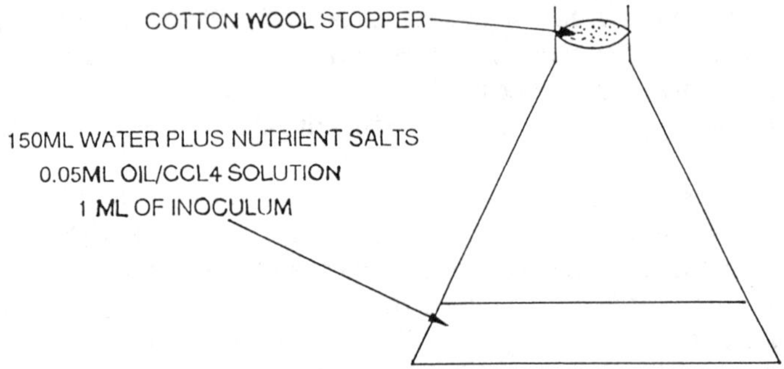


CONDITIONS: STIRRED AT 25`C IN A DARKENED ROOM FOR 21 DAYS

FIG. 2: SCHEMATIC OF THE CEC-L-33-T-82 TEST

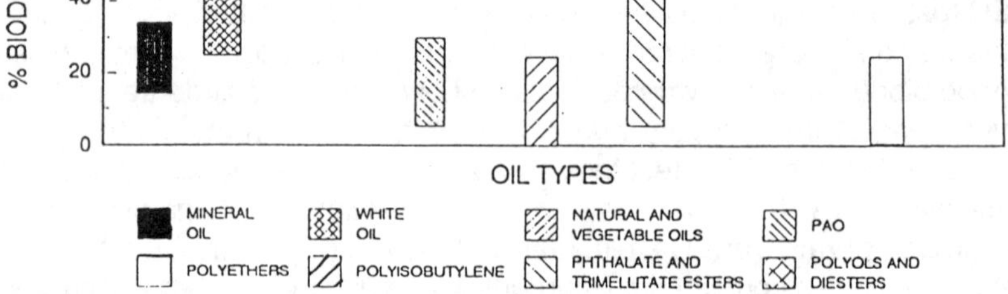


**FIG.3 BIODEGRADABILITY OF VARIOUS LUBRICANTS MEASURED BY CEC-L-33-T-82 TEST**

0 to 80% biodegradable depending on the molecular weight and ethylene oxide/ propylene oxide content. The higher the ethylene oxide content and the lower the propylene oxide content the higher the biodegradability.

The International Council for Marine Industry Association (ICOMIA), a motor industry environmental protection agency, looked intensively into the practical relevance of the CEC test. A motor-boat powered by a two-stroke engine travelled back and forth 600 times within 8 hours in a test tank filled with unadulterated natural mountain stream water: in the first half of the day it used synthetic oil (1:100) in the fuel mix and in the afternoon, mineral oil (1:50). The oil content in the water, determined by Bayerische Landesanstalt für Wasserfürschung (Bavarian State Institute for Water Research), showed that the synthetic oil was degraded immediately and was 95% eliminated at the end of the test phase.[4] Mineral oil biodegraded by only 50%.

3.1.2 The mechanism of biodegradation of ester lubricants. The biochemistry of microbial attack of these esters is known in general outline and has been well reviewed. Ester hydrolysis,[5] ß-oxidation of long chain hydrocarbons[6] and oxygenase attack on aromatic nuclei[7] (the main steps) have been extensively investigated with a variety of micro-organisms. In some cases it has been found that enzymes encoded by plasmid bourne genes play a special role[8]. The main features which slow or reduce microbial breakdown are the extent of branching, which reduces ß-oxidation, and the degree to which ester hydrolysis is inhibited.

Figure 4 shows the results of a series of tests conducted in four different laboratories for ester based lubricants. It is likely that the scatter which occurs between laboratories is the result of the fact that the sewage inoculum can differ from area to area. Also difference in results could be due to differing shake speeds and the degree of temperature control in the incubator.

A particularly high scatter in biodegradabilities is seen for trimellitate esters and further research is ongoing to investigate this. One theory is that the partition coefficient of the oil between the solvent and the aquatic environment has a very significant effect on the biodegradation as it controls the release of the oil into water.

CEC data can be used to identify chemical structure/biodegradability relationships.[9] Figure 5 shows that the biodegradability is reduced with increasing chain length for a range of branched phthalate esters.

3.1.3 Modification of the CEC-L-33-T-82 test. The CEC test gives an indication of the fate of an oil discharged into the domestic sewage system.

| ESTER LUBRICANT | LAB 1 | LAB 2 | LAB 3 | LAB 4 |
|---|---|---|---|---|
| MINERAL OIL | 34 | 40 | 29 | 37 |
| DTDA {DI ESTER} | 92 | 84 | 80 | 88 |
| DIMER ACID ESTER 1 | 18 | 17 | 25 | - |
| DIMER ACID ESTER 2 | 49 | 72 | 50 | 66 |
| DIMER ACID ESTER 3 | 78 | 54 | - | - |
| TMP POLYOL | 96 | 99 | 100 | 100 |
| TRIMELLITATE | 64 | 28 | 16 | - |
| PHTHALATE | - | - | 67 | 70 |

**FIG. 4: A LABORATORY ROUND ROBIN TEST TO EVALUATE THE RELIABILITY OF THE CEC-L-33-T-82 BIODEGRADABILITY TEST**

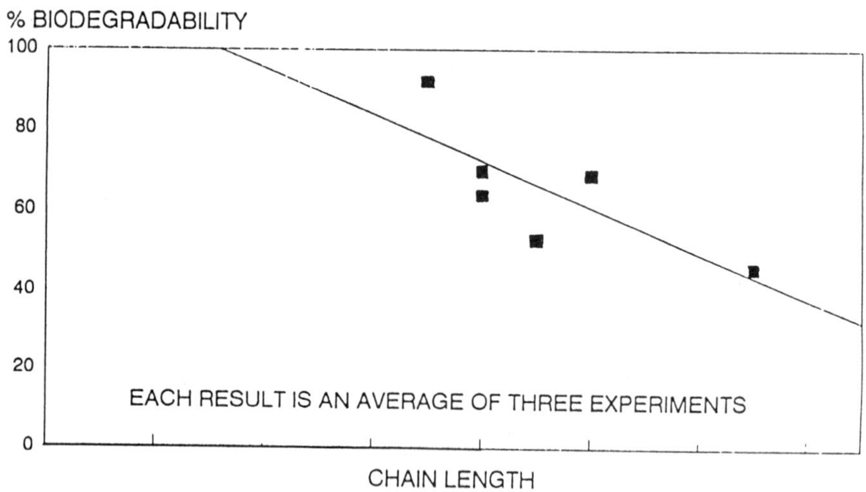


**FIG. 5: A SCHEMATIC OF BIODEGRADABILITY VERSUS CHAIN LENGTH FOR A RANGE OF PHTHALATE ESTERS USING THE CEC-L-33-T-82 TEST.**

Settled sewage contains in the region of $10^6$ c.f.u ml$^{-1}$. This is a much higher cell density than that found in a lake environment ($10^3$ - $10^4$ c.f.u. ml$^{-1}$). We have undertaken experiments in which the sewage inoculum recommended by the CEC test was replaced with river water to give a much better reproduction of the natural environment. River water was collected from the River Tyne at Chollerford Bridge and the natural microbial content concentrated to $10^5$ c.f.u. ml$^{-1}$. This inoculum was then used instead of the sewage inoculum. Results using the modified CEC test are depicted in figure 6. It is interesting to note that the polyol ester biodegrades at the same rate as rape seed oil.

3.1.4 The effect of nutrients on biodegradability. The introduction to the environment of large amounts of oil provides an excellent energy source for microbial growth which results in rapid proliferation of the number of oil-utilising micro-organisms. Nitrogen is needed to sustain this growth as it is a component of protein and nucleic acids being synthesised by the organism. Using the modified CEC test the effects of nitrogen level on the biodegradability of a biodegradable lubricant was investigated (figure 7). The river water used for this series of experiments was concentrated to $10^3$ c.f.u. ml$^{-1}$. Initial degradation in the "no-nitrogen" flasks was probably due to trace nitrogen present in the lake-water inoculum. With increasing nitrogen supplementation, a marked increase in the extent of biodegradation was observed. This trend continued until an optimum carbon to nitrogen ratio, usually about 10:1, was reached, above which increasing nitrogen levels had no effect. Laboratories using inoculum, for the CEC test, with are nitrate rich (from agricultural fertilisers etc.) may therefore show higher values than from laboratories using inoculum from non-agricultural areas. This could account for some of the scatter between laboratories in figure 4. Such variability could be reduced by using suspensions of the microbia inocula, washed twice and concentrated by centrifugation.

3.1.5 Effects of temperature on biodegradability. Laboratory conditions for the CEC test are adjusted for optimum microbial growth with a temperature of 25 C and provision of adequate mineral nutrients in the mineral medium. Temperature is one of the main factors influencing the growth of micro-organisms. Generally, as the temperature rises enzymatic reactions and metabolism proceed at greater rates until a temperature is reached at which proteins become denatured. Generally, an increase in temperature results in an increase in the rate of biodegradation of the oil. For many organisms the optimum growth rate is at 20 to 25 C, when biodegradation can be from twice to ten fold faster than at 4 to 5 C.[10] Since many lakes never reach a temperature of 25 C, field degradation rates would be expected to be lower. Changes can be expected, and have been recorded in degradation rates throughout the year by seasons with little degradation in winter when lake temperatures can reach

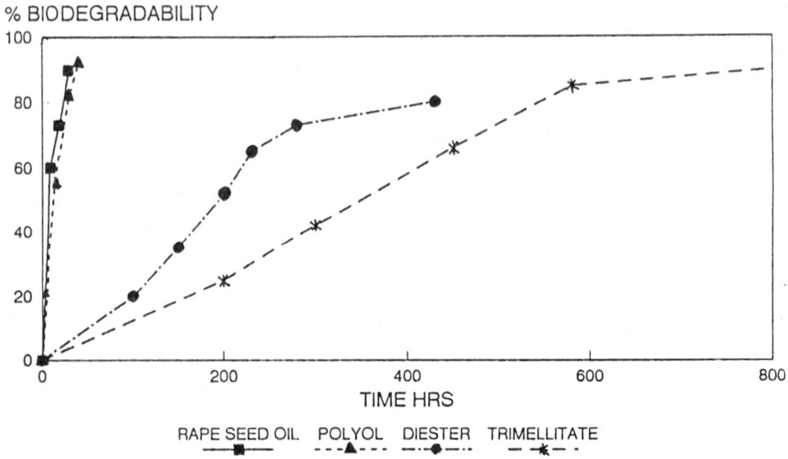


**Fig. 6 BIODEGRADABILITY OF LUBRICANTS USING A MODIFIED CEC-L-33-T-82 TEST.**

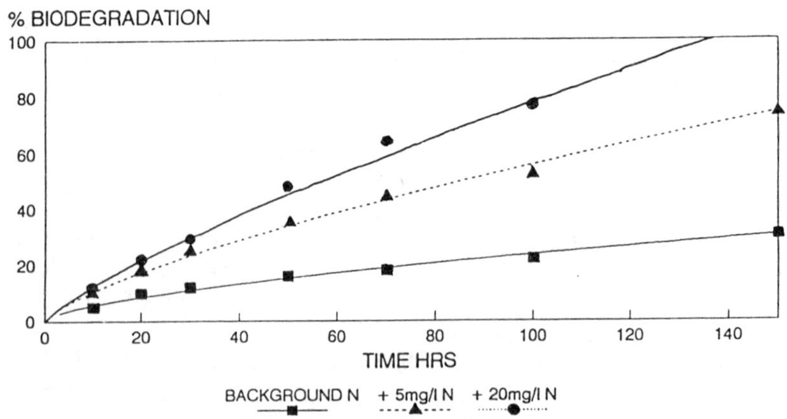


**FIG.7 EFFECT OF EXTRA NITROGEN ON BIODEGRADATION**

0 C, but more in summer when temperatures will be closer to the optimum. This effect may be compensated by seasonal changes in the population of microbial communities (psychrophils) capable of low temperature growth and substrate utilisation. Gilbert and Higgins[11] recorded that in the Medway Estuary of the River Thames, weathered samples of North Sea crude were degraded three times as fast at 12 C than at 5 C. Using the modified CEC test, with the river water inocula concentrated to $10^5$ c.f.u. $ml^{-1}$, a biodegradable lubricant (TMP ester) is shown to degrade three times faster at 25 C than it did at 5 C, see figure 8.

3.1.6 Effects of additives on biodegradability. Fully formulated lubricants often contain additives to boost their physical properties. The presence of additives in the lubricant is known to have a significant effect on the rate of biodegradation although specific knowledge on the individual classes of additives are scarce. In general, additive components which contain transition metals have a very negative effect on biodegradability. It is also likely that certain dispersant/detergent additives could effect the movement of organisms and the transport of nutrients and enzymes across the cell membrane which again will probably adversely effect the rate of biodegradation. On the positive side nitrogen and phosphorus containing additives may be beneficial if the nutrient medium lacks these elements in adequate concentrations to sustain appropriate levels of microbial growth (which is unlikely to be the case in most inland waterways). Work to investigate how additives effect the biodegradability of ester lubricants is presently underway.

### 3.2 Bio-accumulation in the environment

Initial in-vitro predictive tests for the mutagenicity of used engine oil have been determined by means of AMES testing. It was found that 70% of these effects were caused by polyaromatic hydrocarbons (PAHs) with more than three rings, although this fraction represents only 1% of the volume of a used oil.[12] Of this highly mutagenic fraction, 18% of the effect was caused by benz[a]pyrene. According to the International Agency on Research for Cancer, benz[a]pyrene, benz[e]pyrene, benz[a]anthracene and chrysene are considered to have high carcinogenic potential.[13] Graf and Winter[14] found that samples of crude mineral oil from the Persian Gulf, Libya and Venezula were found to contain 400, 1 320 and 1 600 $\mu g\ kg^{-1}$ respectively of benz[a]pyrene.

Oil has a tendency to accumulate in the environment. This can be seen quite clearly in parking lots, in garages and in the central part of highway lanes, especially when there is a dip in the road. When it rains, these PAH-containing oils then reach the sewers, rivers and lakes and finally the sediments. A recent study of Bodensee sediments has clearly shown that these types of hydrocarbons are accumulating near river mouths,[15] see figure 9. There is also a diffuse contribution via rainfall over the whole surface of this lake caused by particulate emissions. The more readily degradable lubricants used in mixed lubricant engines were not found in this study, not even in the entrance to busy marinas, where they could have been expected in measurable concentrations (two-stroke engines emit on average about 25%[1] of their lubricating oil). In another study,[16] bio-assays of the Bodensee sediments confirmed that the environment load of organic compounds mainly originates from rivers. No such loads were found near pleasure boat harbours where there was considerable boat traffic, mainly using mixed lubricant engines.

European regulatory authorities are looking with increasing displeasure at the chemical "soup" of many base oils and favouring high purity synthetics instead. Additives, used to give the neat oil better performance, are coming under attack, most specifically nitrates. The superior performance of lubricants over mineral oil and vegetable oils usually means less additives need to be added to the final formulation. This reduction in additives levels is beneficial as additives often cause bio-accumulation problems.

## 4 TOXICITY

Since the 1970s the need to gain a better understanding of the possible hazards of chemicals in the environment has become widely accepted. As a result, legislation has now been passed in many parts of the world which requires that, before a new substance can be marketed (or manufactured), certain data (which may include ecotoxicological data) must be submitted to the relevant authorities so that the hazard can be assessed. Japan was the first country to pass such legislation in 1973, followed by the implementation of the Toxic Substances Control Act (TSCA) in the USA and passing by the EEC of the sixth amendment in 1979 which ensures there will be similar legislation in EEC countries. Other countries which now have, or will soon have comparable legislation include Australia (Australian Inventory of Chemical Substances, AICS), Canada, Finland, New Zealand, Norway and Switzerland and most other OECD countries are likely to follow suit in due course.

WasserGefahrdungsklasse (WGK, The Water Hazard Classification) is a German model for assessing the effect of functional fluids on the aquatic environment. It uses mammalian acute toxicity, fish toxicity and biodegradability to grade each fluid with a score between 0 and 3. These classifications are explained in figure 10.

In 1989 a substance list concerning classifications into water-endangering classes was published.[17] This list covered:

a) Substances which have been classified by the committee for the evaluation of water-endangering substances (KBWS). Classifications carried out by the KBWS can be recognised by the label “KBWS-Einst” (KBWS-classification).

b) Substances which have been classified by the committee for the evaluation of water-endangering substances themselves but which have not yet been published in the common ministerial gazette. This can be recognised by the capital letters and the indication “KBWS-Einst-V” (which means provisional classification).

c) Substances where the companies of the chemical industry made by the manufacturers themselves according to the VCI concept. This can be recognised by the label “Selbsteneinstufung” (classification by the manufacturer himself).

This information is summarised in figure 11. Ester lubricants can be chosen to give low WGK values.

## 5 EXHAUST EMISSIONS

The hydrocarbon content of diesel exhaust particulates has been found to include a significant contribution from lubricating oil, unchanged by pyrolysis or partial combustion, and which is broadly identical to the engine oil in the sump. Work has shown that the lubricant contribution is within the range 40 to 65% of the solvent soluble organic fraction (SOF).[18] This unchanged lubricating oil arises from leakages, via the valve stem seals or piston rings, during periods other than the combustion part of the engine cycle. Ester lubricants can be tailor-made to reduce the amounts of smoke emitted in the exhaust gases.

Abas et al[19] have shown that 40 to 70% of the PAH in exhaust particulates stem from the PAH accumulated in the engine oil. The other 30 to 60% originate from the fuel or the combustion process. A decrease of 25% of total PAH content has been found by using a carboxylic ester over a mineral oil in a Vespa single cylinder, two-stroke, air cooled motor run at constant power output of 3 horse power at a speed of 4 300 revolutions per minute.[13] PAH emission from the ester run experiment were found not to be appreciably affected by a temperature increase up to 100 C, nor by an increase of lubricant concentration from 2 to 4% in the fuel mixture.

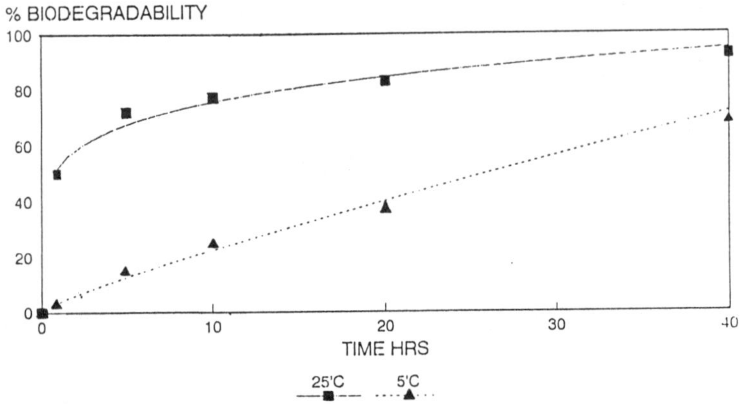


FIG. 8: THE EFFECTS OF TEMPERATURE ON BIODEGRADABILITY USING A MODIFIED CEC-L-33-T-82 TEST

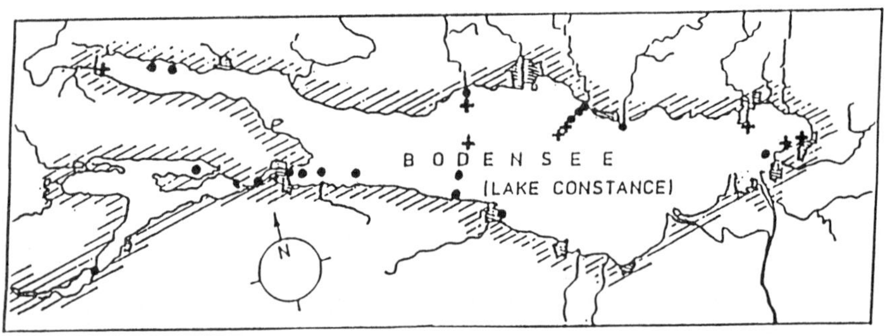


+ HIGH CONCENTRATIONS

**FIG. 9 SURVEY OF PAH LEVELS ON THE BODENSEE**

- 0 - DOES NOT GENERALLY ENDANGER WATER QUALITY
- 1 - SLIGHTLY ENDANGERS WATER QUALITY
- 2 - ENDANGERS WATER QUALITY
- 3 - SERIOUSLY ENDANGERS WATER QUALITY

FIGURE 10: WATER HAZARD CLASSIFICATION WGK (WASSERGEFAHRDUNGKLASSE)

| LUBRICANT | WGK VALUE |
|---|---|
| PAGs | 0 TO 1 |
| ESTERS | 0 TO 2 |
| PAOs | 1 TO 2 |
| MINERAL OILS | 2 |

FIG. 11: A SUMMARY OF WGK VALUES FOR VARIOUS TYPES OF LUBRICANTS.

## 6 PERFORMANCE OF A LUBRICANT

Synthetic lubricants can provide superior performance to traditional mineral based oils. This is especially true as synthetics give the chemist the opportunity to tailor-make the lubricant for the application. Their superior lubricating properties, for instance, often allows them to be used in smaller quantities or to be used more efficiently. For example, esters can be used at dilution ratios of 100:1 in marine two-stroke oils compared with 50:1 for mineral oil. Also, because of their low volatility and increased thermal stabilities they often have longer service lives. The advantage of using less lubricant for longer periods of course benefits the environment. Performance of an environmentally friendly fluid is important as various physical property specifications have to be met depending on the application. The main drawback of vegetable oil alternatives to synthetic esters is that they often lack the performance needed for demanding applications. Rape seed oils have been shown to exhibit half to a quarter of the life of synthetics in particular applications.

## 7 CONCLUSION

Synthetic lubricants, especially esters, offer opportunities for meeting the high performance requirements of today's lubricants whilst reducing deleterious effects on the environment. As we become more concerned with the environment no doubt the use of synthetics will increase.

## REFERENCES


1. P. Van Donkelaar, *The Science of the Total Environment,* 1990, **92**, 165.
2. W. Kuhrt and T.H. Mang, "Environmental acceptable Lubricants and legislation", Deutsches Scierstoff-Forum, Frankfurt, November 1990.
3. T. Henderson, "Simple ways to help a planet in dire peril", *The Journal* (a Cleveland newspaper), Monday 11th June 1990.
4. B. Wachs and H. Wagner, *Erdol und Kohle*, 1982, **35**(5), 186.
5. A.R. Macrae and R.C. Hammond, *Biotechnology and Genetic Engineering Reviews*, **3**, 1093, 217
6. J.M. Wyatt, Ph D Thesis, Unversity of Kent, 1982.
7. C.E. Cerniglia, 'Petroleum Microbiology', ed R.M. Atlas, Macmillan, New York, 99.
8. C.E. Chakrabarty, 'Biotechnology', ed H.J. Rehn and G. Read, **8**, 515.
9. M. Wright, Cranfield Institute of Technology, unpublished work.
10. R. Atlas, unpublished work, 1984.
11. P.D. Gilbert and I.J. Higgins, *J. of General Microbiology*, 1978, **108**, 63.
12. G. Grimmer et al, *Erdol und Kohle*, 1982, **35**(10), 466.
13. E. Cosmacini, D. Cottia, L. Pozzoli and R. Leoni, *J. of Synthetic Lubricants*, 1988, **3**, 251.
14. W. Graf and C. Winter, *Arch. Hyg. Bakt.*, 1968, **152**, 289.

15. S,E. Howells et al, 'Analysis of hydrocarbons in sediments of the Bodensee (Lake Constance)', Oil Pollution Research Unit', Pembrook, 1986, report no. FSC/OPRU/3/86.
16. Internationale Gewasserschutzkommission für den Bodensee, 'Die Ologochaeten im Bodensee als Indikatoren für die Belastung des Seebodens (1972 bis 1978), Bericht 38, Bodensee, West Germany, 1988.
17. Substances list concerning WGK, issued by KBWS, March 1989.
18. C.J. Trier, G.S. Petch, M.M. Rhead, D.E. Fussey and P.R. Shore, 'Assessing the contribution of the lubricating oil to diesel exhaust particulates with particular reference to polycyclic aromatic hydrocarbons', Proceedings of the Institute of Mechanical Engineers, Tribology-Friction, Lubrication and Wear Fifty Years On, 1987, C340.
19. M.K. Abas et al, 'The ageing of lubricating oil, the influence of unburnt fuel and particulate SOF contamination', Society of Automotive Engineers, Nov. 1987, paper 872085.

# NEW AUTOMOTIVE TECHNOLOGY WITH POLYURETHANE FOR INTERIOR TRIM;

## INTEGRATED SEATING OR DOOR PANELS.

G. Jackson

BASF Elastogran
Somercotes
Derby, DE55 4NL
United Kingdom

## 1 INTRODUCTION

Polyurethane polymers are characterised by the presence of the urethane group or linkage in the main polyurethane chain -NH-CO-O-. However, in many of today's commonly used polyurethanes other groups such as amide, ester and ether may predominate. Consequently, the term polyurethane is used to describe the many and varied complex reaction products of polyisocyanates and polyhydroxy compounds.

## 2 RAW MATERIALS

The urethane group or linkage is formed by the reaction of an isocyanate and a hydroxyl containing compound.

| R-NCO | + | R'-OH | R-N-C=O<br>H O-R' |
|---|---|---|---|
| isocyanate | | hydroxy compound | urethane |

The use of multi-functional reactants, usually di-isocyanates and polyhydric compounds (polyols) will lead to the formation of a cross linked polymer; a polyurethane.

| OCN-R-NCO | + | HO-R'-OH | OCN-R-NHCO-R'-OCNH-R-NCO (C=O, C=O) |
|---|---|---|---|
| di-isocyanate | | polyol | polyurethane |

Therefore, the important raw materials are di-isocyanates, diols, triols and polyols. Additives such as catalysts, blowing agents and cell control agents will be discussed later.

## <u>Isocyanates</u>

In commercially useful polyurethanes di-isocyanates are most commonly used in the manufacture of polyurethanes. The commercially important isocyanates are:

<u>80/20 Toluene Di-isocyanate (80/20 TDI)</u>. This is an 80/20 mixture of toluene-2,4-di-isocyanate and toluene-2,6-di-isocyanate and is used mainly in the manufacture of flexible slabstock and moulded polyurethane foam.

<u>65/35 Toluene Di-isocyanate (65/35 TDI)</u>. This is a 65/35 mixture of toluene-2,4-di-isocyanate and toluene-2,6-di-isocyanate and is used mainly in flexible polyester slabstock foam although not to as great an extent as 80/20 TDI.

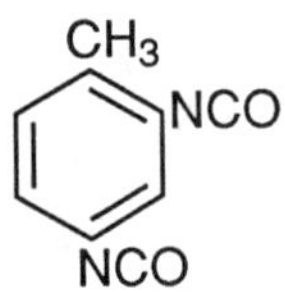


2,4-toluene-di-isocyanate

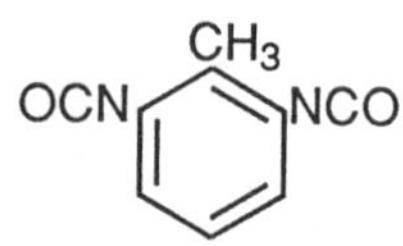


2,6-toluene-di-isocyanate

<u>Di-phenylmethyl Di-isocyanate (MDI)</u>. During production of MDI, aniline is reacted with formaldehyde. The product formed consists mainly of 4,4-di-amino-diphenylmethane with smaller amounts of the 2,4 isomer and polyamines. The amine mixture is then phosgenated without separation to give 'crude' diphenylmethane-di-isocyanate which typically contains the following:

55% diphenylmethane-di-isocyanate (4,4 and 2,4 isomers),
25% tri-isocyanates,
20% higher polyisocyanates.

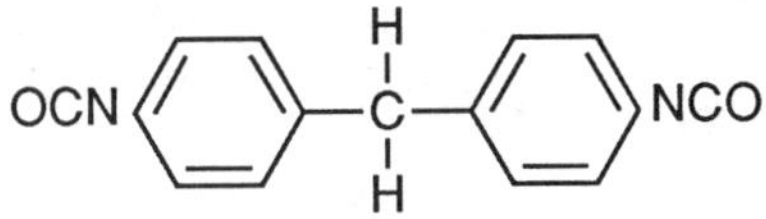


Diphenylmethane-4,4-di-isocyanate

Diphenylmethane-di-isocyanate is a solid with a melting point of 37 to 38 C, whereas 'crude' MDI is a liquid at room temperature. Both products have significantly lower vapour pressure than toluene-di-isocyanate and are therefore less hazardous in use. Various degrees of refinement of 'crude' MDI leads to a wide variety of semi-refined MDIs which can be used to enhance the various properties of the derived polyurethane.

Polyols. The earliest polyurethanes were based on aliphatic diols such as butan-1,4-diol. However, it is now more usual for polyurethanes to be produced from polymeric hydroxy compounds, and of these polyester and polyether polyols are the most commonly used.

Polyether Polyols. In commercial polyurethane production hydroxy terminated polyether polyols have attained a dominant role. The most widely used polyether polyols are derivatives of propylene oxide and ethylene oxide. Today it is possible to produce highly reactive polyether polyols. Normally triols are produced by polymerising propylene oxide in the presence of a trihydroxy compound such as glycerol. Ethylene oxide increases the reactivity of the polyol, however, the greatest increase in reactivity is obtained by reacting the ethylene oxide onto the end of the polyol chain, thereby producing a material with a high percentage of primary hydroxyl groups which are highly reactive in urethane formation.

Obviously, highly reactive components are desirable if rapid curing of the polyurethane polymer (and therefore quick demould and cycle times) are to be attained. Polyols for conventional slabstock production have relatively low levels of primary hydroxyl groups and molecular weight ranges of 3 000 to 3 500. Highly reactive polyols for use in moulded polyurethane foam elastomers and coatings have a high level of primary hydroxyl groups and a molecular weight range of 4 800 to 6 000. Rigid polyurethane foams are produced from triols of low molecular weight (about 500) or from polyols of higher functionality, prepared by polymerising propylene oxide in the presence of hydroxy compounds such as sorbitol.

Crosslinking Agents. The crosslinking agents most commonly used are low molecular weight polyols, diols, triols, glycerol and di- and tri-ethanolamines.

Catalysts. Catalysts are usually amine or tin based compounds. They accelerate and control the water/isocyanate reaction and polyol/isocyanate reactions.

Silicone Surfactants. Silicone surfactants provide two essential functions:

a) to assist the mixing of the two liquid components to form a homogeneous liquid,
b) to stabilise the gas bubbles during foam expansion, so preventing collapse before the liquid polymerises, that is control of cell size and structure.

Blowing Agents. The primary blowing agent which causes the foam to expand is carbon dioxide which is generated from the reaction between water and isocyanate.

$$\underset{\text{isocyanate}}{\text{R-NCO}} + \underset{\text{water}}{H_2O} \qquad \underset{\text{amine}}{\text{R-NH}_2} + \underset{\text{carbon dioxide}}{CO_2}$$

A secondary blowing agent may be used in addition to water in order to modify the properties and/or nature of the foam. These secondary blowing agents are low boiling point liquids such as trichlorofluoromethane (CFC-11), boiling point 23 C. Since some of these compounds degrade ozone in the upper atmosphere, replacements are being sought. In all of the polyurethane systems being used in the production of automotive door panels CFCs are not used.

## 3 FOAM FORMATION

Polyurethane foam formation is the result of a series of varied and complex reactions, the two most important reactions being:

a) the reaction between di-isocyanate and polyol,
b) the reaction between di-isocyanate and water.

These reactions by themselves are far too slow for the commercial manufacture of polyurethanes and therefore catalysts are used to increase the rate of reaction and establish the required balance between the polymerisation or chain extension reaction and the foaming reactions.

For all blown polyurethanes the balance between the generation of carbon dioxide and polymer growth is quite critical. The polmer must develop sufficient strength to entrap the generated carbon dioxide, for foaming to take place, and at the end of reaction maintain the structure so as to prevent foam collapse. Catalysts are also essential for controlling the cure rate of the foam and the development of the foam's final physical properties.

During reaction of the two components heat is generated which aids the expansion of the carbon dioxide generated and/or volatilises any low boiling point blowing agent. Gas bubbles form within the mixture and an expanding foam is produced in which simultaneously there is an increase in temperature, volume, viscosity and internal pressure. The reaction profile of a foam system can be characterised by measuring the cream time, fibre time, rise time and free rise density. In order to produce a polyurethane foam the two components have to be intimately mixed in a pre-determined ratio.

A-Component. This usually consists of mainly polyol into which auxiliary components have been pre-blended, so that the A-component may consist of some or all of the following:

a) polyol,
b) cross-linkers,
c) water/blowing agents,
d) catalysts,
e) silicones,
f) pigments, fire retardants, UV stabiliser.

B-Component. Usually this consists solely of the di-isocyanate. The mixing ratio is determined by the ingredients in the A-component, the desired properties of the derived polyurethane and the index.

## 4 DEFINITION OF TERMS

Mixing Ratio. For each system the recommended mixing ratio will be given in the technical data sheets. It indicates the correct amount of isocyanate to be mixed with a given amount of polyol by weight, for example:

| | | |
|---|---|---|
| mixing ratio | A-component/B-component | = 100:40 |
| | A-component/B-component | = 100:40 |
| | A-component/B-component | = 2.5:1.0 |

Cream Time (CT). The time between the start of mixing and the beginning of foam rise. The beginning of foam rise is usually accompanied by the liquid mixture becoming 'creamy' in appearance.

String Time (ST) or Fibre Time (FT). The time between the start of mixing and the point at which the foam has developed gel strength, that is internal strength, to give a stringing of the foam when a plastic straw is inserted and removed from the foam.

Rise Time (RT). The time between the start of mixing and the point where foam rise is complete.

Density. Measurement of the foam's free rise density (FRD) can be easily achieved using foam samples prepared in a cup. The foam is allowed to cure for five minutes and using a sharp knife cut off the excess foam level with the top of the cup and weigh.

$$\text{Free rise density (kg m}^{-3}\text{)} = \frac{\text{foam weight (kg)}}{\text{cup volume (m}^{3}\text{)}}$$

# CHEMICAL ANALYSIS USING FTIR SPECTROSCOPY IN THE AUTOMOTIVE INDUSTRY

M.B. Simpson

Mattson Instruments Ltd
Newport Pagnell
Buckinghamshire, MK16 0AL
United Kingdom

## 1 INTRODUCTION

The enormously enhanced speed and sensitivity of Fourier-transform infra red (FTIR) spectroscopy as compared with previous methods has been common place for some ten years, but much more recently a variety of changes and refinements in the available instrumentation have combined to make these advantages routinely accessible in demanding industrial circumstances. The automotive industry is really no different (except perhaps in the increasingly over-riding significance of its quality systems) from any other engineering sector industrial user of FTIR. The principal requirement is for the rapid analysis of difficult samples by direct means (that is with the minimum of sample preparation), followed by an informative analysis of the data without the need for extensive spectral interpretation. As an example, the analytical requirement for condition monitoring of used engine oil might easily be over 100 samples per day. Quite apart from the obvious need for a rapid sampling technique, it is clear that detailed individual spectral analysis cannot be undertaken in such a regime, and must be replaced with a quantitative semi-expert system approach. To summarise, the major applications of FTIR spectroscopy to automotive industry chemicals include the analysis of difficult samples in the raw state, semi-automatic quality control and monitoring in use, creation of user-databases to analyse different formulation recipes, and finally as one tool in a combined multi-analytical approach to failure analysis.

These aspects of FTIR spectroscopy in the context of the automotive industry will be illustrated with four examples covering multi-component partial least squares regression (PLS) calibration of additive packages in commercial motor oil, condition monitoring of used oil, quality control (QC) discrimination between different formulations of a complex resin-based mixture, and routine analysis of automobile plastics for failure analysis. All spectrscopic data presented in the examples shown here were obtained using a Mattson Instruments GALAXY 4020 spectrometer fitted with a DTGS detector, running FIRST software under PC-AT control.

## 2 APPLICATIONS OF FTIR SPECTROSCOPY

The main feature of all these examples is that the form of the sample as presented to the instrument for analysis is not at all suitable for routine transmission spectroscopy, which would, of course, have been the only method feasible with dispersive IR instruments. Most real-life samples are much too intensely absorbing or scattering for this to be possible. Which is not to say that their routine measurement is impossible; on the contrary,the use of sampling procedures such as attenuated total reflectance (ATR) and diffuse reflectance (DRIFTS) enable the spectra of oils, plastics and resins to be measured on a routine basis. The first two examples to be described are both problems associated with oil analysis, where ATR techniques work very well, whereas the second two examples are solid polymer or resin samples where DRIFTS is particularly applicable.

### Multi-component PLS Calibration of Oil-additives

Most lubricating oil formulations contain in addition to the base-oil blend an additive package consisting of a number of components designed to inhibit oxidative degradation, aid emulsification and suspension of particulate contamination, as well as helping to maintain the oil's viscosity. A typical requirement might be to provide a quantitative analysis of such additives, present as a two or three-component mixture at the 5 to 10% level, with an accuracy of about 0.1%. This might be achieved using capillary gas chromatographic methods, which will, however, be quite slow. Since the additives are present at low level their spectral contribution is small and overlap both with each other and the base-oil. This situation is ideal for the application of the computational technique of PLS for quantitative analysis.[1] Once calibrated using PLS multivarate techniques, FTIR analysis offers an almost instantaneous analysis of many-component mixtures.

Thus figure 1 compares the spectra, obtained with one minute data aquisition using an hATR accessory, of the base-oil blend (b) and a typical example of the formulated product (a) containing two additives. In both cases the spectra are dominated by the absorptions of the base-oil, but in (a) the weakly visible contributions from the additives are arrowed. Figure 2 expands the region between 1800 - 1500 $cm^{-1}$ showing the overlapping absorptions arising from the base-oil, and the two additives. Rather than attempt to isolate any individual spectral features the PLS multivariate calibration routine uses *all* the spectral information in this selected window, automatically optimises an N-dimensional fit of the calibration data to the given additive concentrations and generates a calibration matrix, which can then be used to predict the concentration of these additives in unknown mixtures. An important feature of the PLS method is that it allows for validation of the calibration data-set, by sequentially removing one standard at a time and, after recalibrating, predicting the known composition of that standard, which provides a trap for outlier detection, and also gives an estimate of the calibration error. The result of this procedure is shown in table 1, where it is clear that the additive composition may be determined to within 0.1% by weight absolute. In this particular example dimensionalities (optimum numbers of factors) of 5 and 8 were used for additives A and B respectively. The PLS technique checks the total error associated with the use of a given number of factors and also enables the vector pairs used for each factor to be examined, thus allowing the best fit to be obtained for the calibration matrix.

## Condition-monitoring of Used Engine Oil

This example continues the theme of oil analysis, but instead of controlling the formulation of oil blends it illustrates the more difficult quantitative problem of examining the degradation of used oils.[2] Typically this includes loading of the oil with particulate carbon, breakdown of the oil into products of oxidation and nitration, contamination with water or glycol, or possibly dilution with unburnt fuel. All of these may be detected easily by spectroscopic methods, although their quantitative assay presents problems of varying difficulty. Perhaps the key analysis of used oil is to determine its carbon loading, which is traditionally undertaken using an extremely laborious procedure involving solvent extraction and centrifugal separation. The analysis is of extreme commercial significance, since modern lubricants for large engines can be designed to run for the entire life of the engine without change.

Figure 3 compares two hATR spectra obtained for unused and used oil. The effect of the carbon particles in the used is partly to absorb, but mostly to scatter (depending on the particle size) the infra-red beam, producing an offset to the

**Table 1**

Cross-validation matrix, showing the predictive self-consistency of the calibration data-set. Such information is vital as a check on PLS procedures, and to detect any outliers among the standards.

| Standard | Component | Predict Conc | Actual Conc |
|---|---|---|---|
| mo2 | additive A | +1.1 | +1.0 |
| | additive B | +14.8 | +14.8 |
| | base oil | +83.9 | +84.2 |
| mo3 | additive A | +2.2 | +2.1 |
| | additive B | +13.9 | +14.0 |
| | base oil | +83.9 | +83.9 |
| mo4 | additive A | +2.9 | +3.0 |
| | additive B | +4.1 | +4.1 |
| | base oil | +92.9 | +92.9 |
| mo5 | additive A | +4.0 | +4.0 |
| | additive B | +10.0 | +10.0 |
| | base oil | +86.0 | +86.0 |
| mo6 | additive A | +4.8 | +4.9 |
| | additive B | +2.1 | +2.0 |
| | base oil | +93.2 | +93.1 |
| mo7 | additive A | +6.0 | +5.9 |
| | additive B | +8.0 | +8.0 |
| | base oil | +86.0 | +86.1 |
| mo8 | additive A | +6.8 | +6.8 |
| | additive B | +12.1 | +12.0 |
| | base oil | +81.2 | +81.2 |
| mo9 | additive A | +8.0 | +8.0 |
| | additive B | +5.9 | +6.0 |
| | base oil | +85.9 | +86.0 |
| mo10 | additive A | +9.2 | +9.2 |
| | additive B | +11.2 | +11.2 |
| | base oil | +79.7 | +79.6 |
| mo11 | additive A | +10.0 | +10.0 |
| | additive B | +12.7 | +12.7 |
| | base oil | +77.3 | +77.3 |
| mo12 | additive A | +9.9 | +10.0 |
| | additive B | -0.2 | +0.0 |
| | base oil | +90.2 | +90.0 |
| mo13 | additive A | -0.1 | +0.0 |
| | additive B | +15.1 | +15.0 |
| | base oil | +85.1 | +85.0 |

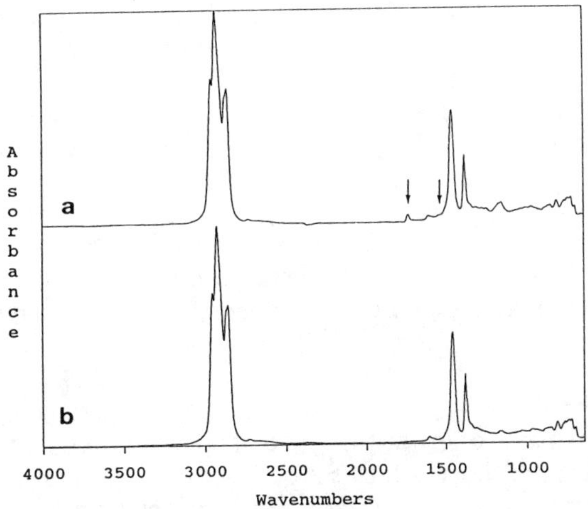


**Figure 1**

(a) Oil blend with additive pack, measured using hATR, absorption bands due to two-component additive mixture arrowed.

(b) Base-oil blend.

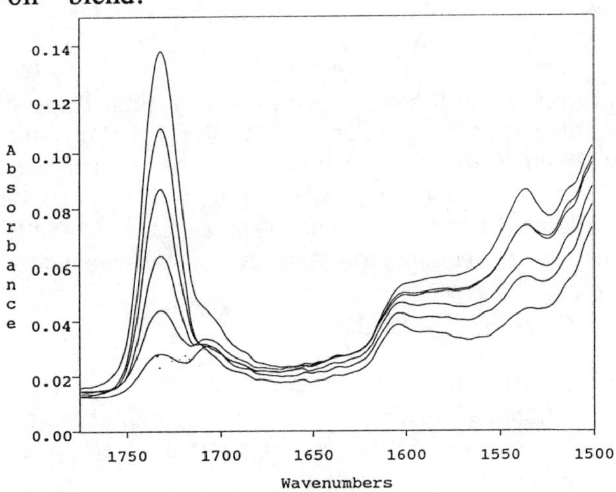


**Figure 2**

Expanded window showing region of maximum variance in the spectra of calibration standard mixtures of the type shown in Figure 1.

spectral baseline. The extent of this baseline offset is quantifiable, and may be calibrated against carbon loading measured using the alternative wet-chemical procedure. A typical calibration curve for baseline absorbance offset verses carbon loading for a sample set with carbon loadings in the range 0.5 to 5% by weight is shown in figure 4. The exact form of the relationship will be sample-type dependent, and will also be effected by the precise sampling geometry of the hATR accessory employed. Thus, as in almost all quantitative infra-red spectroscopy, the system must be calibrated using a suitable and representative calibration set. Once this is achieved, however, very good quantitative precision is possible over the range of carbon content which is of actual interest in real used-oil samples.

Other examples of degradaton that may be at least semi-quantitatively determined are oxidation/nitration, glycol contamination and fuel-dilution. Figure 5 illustrates the spectral changes consequent on oxidation and nitration, which are much more easily seen in the subtraction spectrum shown in the lower part of the figure. The exact band positions are variable according as the formation of RR'C=O, RCOOH, $RNO_2$ or $RNO_3$ groups predominate. Both glycol and water contamination cause a broad O-H absorption at 3400 $cm^{-1}$, although the glycol also gives rise to a characteristic doublet at 1100 $cm^{-1}$, as shown in figure 6. Finally the effect of fuel dilution is indicated in figure 7, where the characteristic absorption bands arising from the aromatic components of the gasoline appear below 800 $cm^{-1}$.

Somewhat surprisingly, therefore, all the main aspects of used-oil degradation: carbon-loading, oxidation/nitration, water/glycol contamination and fuel dilution all reveal themselves clearly in the infra-red spectrum of the used oil, and moreover do so in a way that enables the extent of the degradation to be quantified. In order to be of practical use this analysis is automated, because once again individual spectral interpretation is too time-consuming.

## QC Control of Formulation Consistency

A common problem in manufacturing is to determine and control the consistency of the formulation of a mixture of raw materials before processing. This falls well short of a requirement for full quantitative analysis; it is simply sufficient to determine that the formulation reproduces an expectation, although that may require very precise differentiation between closely similar mixtures. The example we shall examine is the mixture of modified phenolic resins and fillers used to manufacture brake-lining materials. For these mixtures DRIFTS offers the minimum possible degree of sample preparation, and for such strongly absorbing and scattering materials is capable of producing acceptable spectra

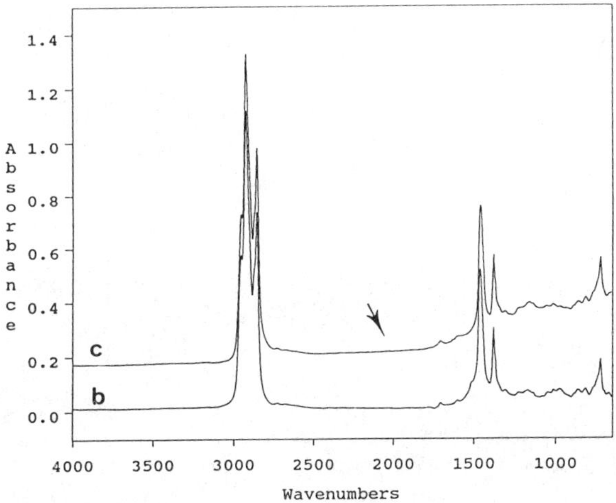


**Figure 3**

Spectra of unused oil (b) and carbon-loaded used oil (c) compared on th same vertical scale. Note the absorbance offset of the baseline (arrowed produced by the soot content of about 4% by weight.

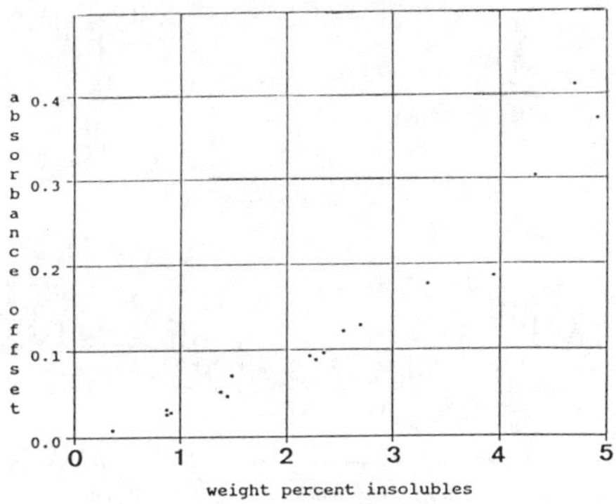


**Figure 4**

Scatter plot of data from seventeen samples of used oil with carbon loadings in the range 0.5-5% by weight. The exact form of the fit will vary according as different sampling procedures are used, but once calibrated for a particular type of oil will yield a rapid and precise analysis.

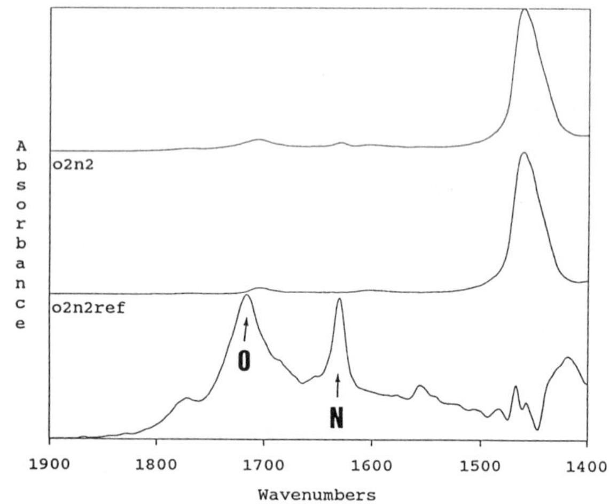


**Figure 5**

Used oil degradation; characteristic features due to oxidation (O) and nitration (N) between 1600 and 1740cm $^{-1}$.

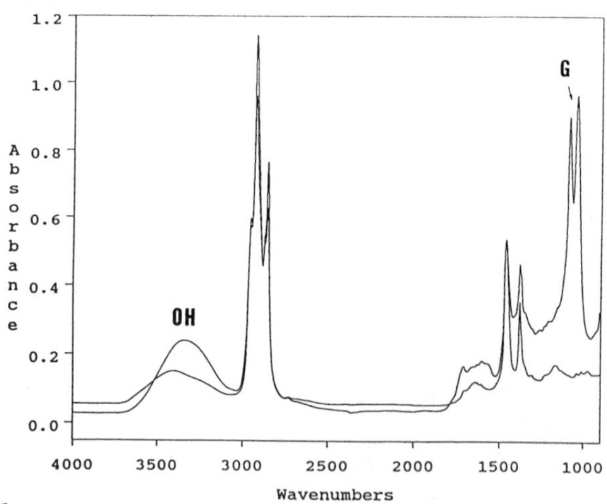


**Figure 6**

Used oil contamination; characteristic features due to water (OH) and glycol (G) contaminants.

in the most reproducible way.[3] It is important, however, when using DRIFTS to ensure reasonable consistency in the particle size distribution of the samples presented for analysis. This is especially so for quantitative work, where dilution with a non-absorbing matrix such as potassium bromide is also necessary, but for this QC task direct measurement of the raw material is possible. For the DRIFTS measurements shown here all the samples were pre-milled to a constant particle size.

The base material for brakepads and linings is a phenolic resin. Small spectroscopic differences show up between the different resins depending on whether oil, ester or nitrile modifiers have been used. Once formulated some mixtures, as in figure 8, are easy to distinguish; the major differences arise from the resin-modifiers and inorganic fillers which are arrowed. However, in a typical formulation list of some fifty recipes many mixtures will be quite similar, figure 9. Again, individual spectral analysis is too slow and unreliable a method to discriminate between such mixtures; library search routines applied to a user-created data base of the formulations are, on the other hand, remarkably quick and accurate in their ability to discriminate and assign confidence limits to the match. An example of this is shown in figure 10 which illustrates the result of searching a brakepad formulation against a database containing the standard reference spectra of about fifty mixtures. The mixture is correctly identified and separated from three other closely similar formulations.

## *In-situ* Sampling and Failure Analysis

Finally we shall illustrate the use of FTIR spectroscopy as an *in-situ* sampling method for failure analysis. Once again DRIFTS is an extremely convenient and rapid technique, especially for polymer samples, when used in conjunction with carborundum or diamond abrasive paper sampling. For this an abrasive pad is used to measure a background spectrum; the pad is then lightly abraded over the polymer surface and the diffuse reflectance spectrum of the sample measured. Figure 11 shows four DRIFTS spectra recorded in this way from different locations on a real vehicle. The measurement time in each case was less than 30 seconds! Such spectra are of more than adequate quality for library searching and identification. Figure12 shows the same exercise repeated on the inner and outer surfaces of a fragment of degraded polyvinylchloride. The oxidative degradation leading to discoloration and cracking of the outer skin is clearly revealed in the carbonyl absorption at 1740 $cm^{-1}$.

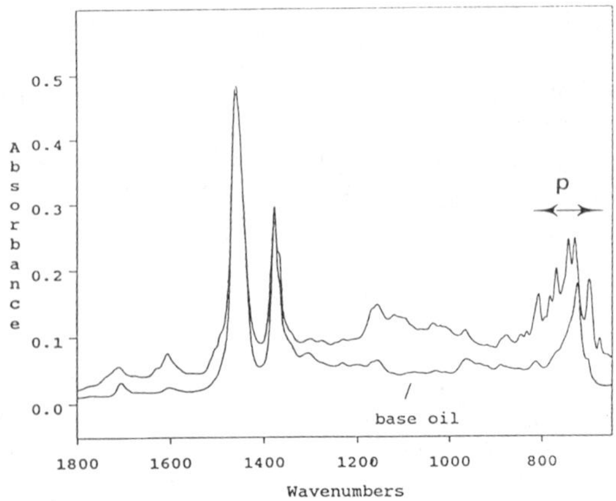


**Figure 7**

Used oil contamination; characteristic features due to fuel dilution and consequent absorption below 800cm$^{-1}$ arising from aromatic gasoline components.

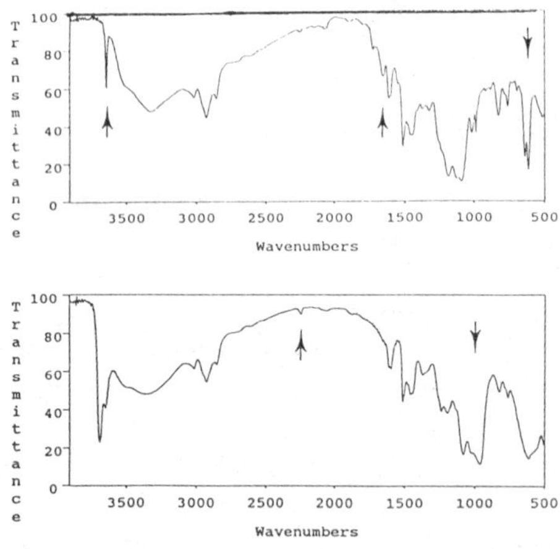


**Figure 8**

DRIFTS spectra of two examples of phenolic-resin based raw materials; the principal differences (arrowed) arise from differing modifiers and inorganic fillers. In this example the formulations are easily distinguished.

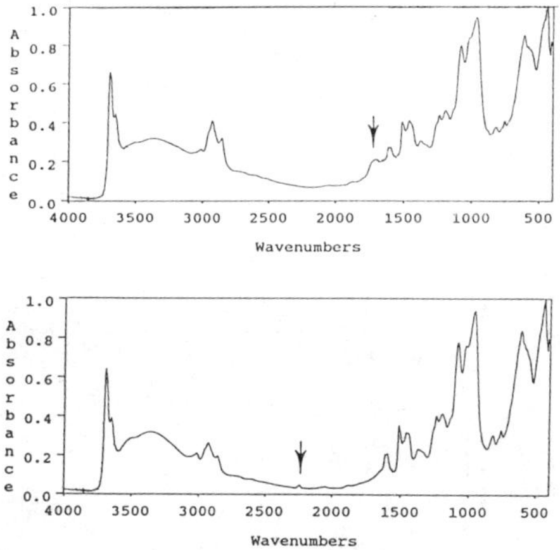


**Figure 9**

Two similar formulations showing only very slight differences (arrowed); database searching is a useful procedure for discriminating among such formulations without the need for individual spectral analysis.

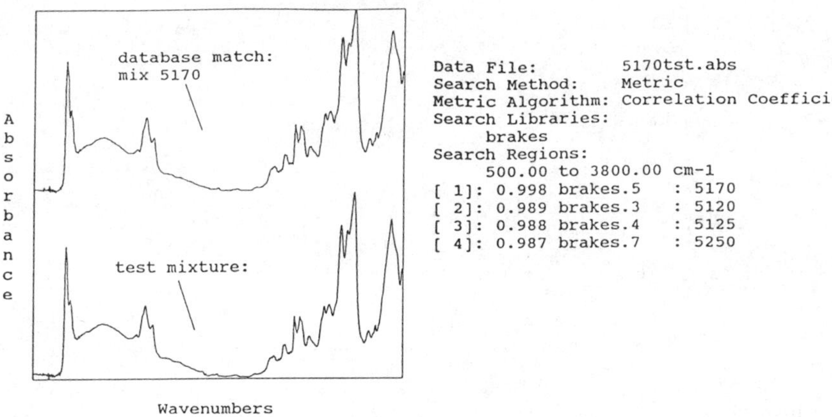


**Figure 10**

Example of database searching based on an assembly of about fifty recipes, showing high level of discrimination.

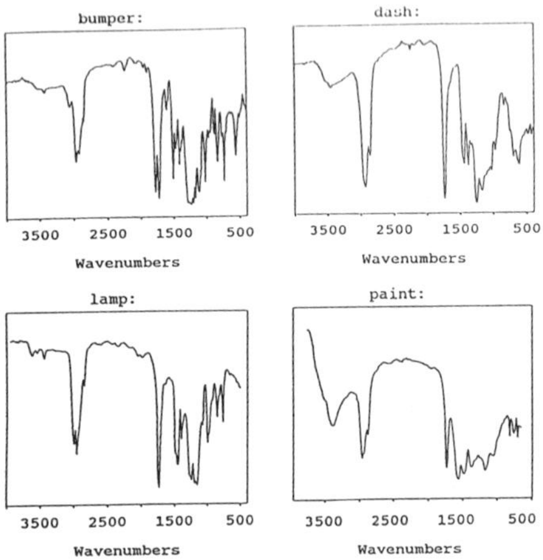


**Figure 11**

Examples of *in-situ* sampling of in-car plastics using the abrasive paper / DRIFTS technique. Data acquisition times were less than 30 seconds for each spectrum.

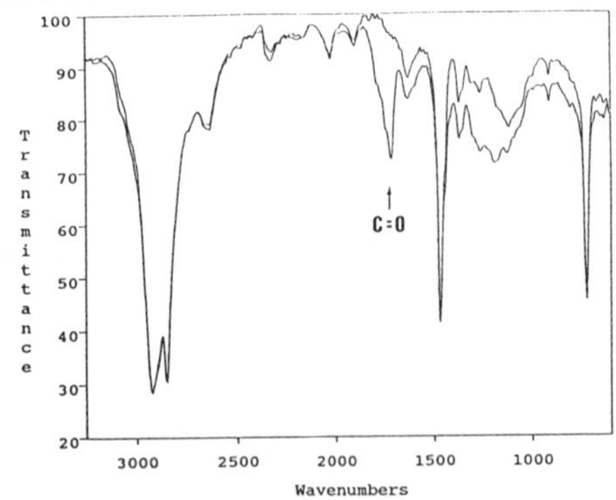


**Figure 12**

Abrasive disc / DRIFTS sampling of inner and outer surfaces of degraded polymer moulding. Oxidative degradation at the surface is clearly evidenced by the C=O absorption at 1740cm $^{-1}$.

## 3 CONCLUSION

In this paper we have illustrated the application of FTIR spectroscopy to some typical problems encountered in the automotive industry. In common with many industrial analytical problems the samples available are not necessarily in the most suitable form for infra-red analysis. However, the choice of an appropriate sampling technique, usually hATR or DRIFTS, is often sufficient to overcome this difficulty. Furthermore, the availability of semi-expert software systems for quantitative analysis and qualitative discrimination can solve many seemingly intractable spectroscopic problems.

## REFERENCES

1. H. Martens and T. Naes, 'Multivariate Calibration', Wiley & Sons, 1989.
2. D.L. Wooton, B.J. Lawrence, J.G. Damrath, SAE Transactions, Paper 841372.
3. J. Chalmers, in 'Advances in Applied FT-IR Spectroscopy', M. W. Mackenzie ed, Wiley, 1988.

SUBJECT INDEX